PRACTICAL ELECTRONIC FAULT FINDING AND TROUBLESHOOTING

ROBIN PAIN

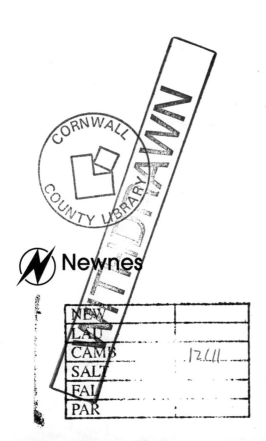

 **Newnes**

Newnes
An imprint of Butterworth-Heinemann
Linacre House, Jordan Hill, Oxford OX2 8DP
A division of Reed Educational and Professional Publishing Ltd

⟨R A member of the Reed Elsevier plc group

OXFORD   BOSTON   JOHANNESBURG
MELBOURNE   NEW DELHI   SINGAPORE

First published 1996
Reprinted 1997
© Reed Educational and Professional Publishing Ltd 1996

British Library Cataloguing in Publication Data
Pain, Robin
    Practical Electronic Fault Finding and Troubleshooting
    I. Title
    621.381

ISBN 0 7506 2461 2

Library of Congress Cataloguing in Publication Data
Pain, Robin
    Practical Electronic Fault Finding and Troubleshooting/Robin Pain
    Includes index.
    ISBN 0 7506 2461 2
    1. Electronic apparatus and appliances – Maintenance and repair.
    2. Electronic apparatus and appliances – Testing. 3. Electronic
    circuits – Testing.   I. Title.
    TK7870.2.P35   95-18106
    621.3815'48–dc20   CIP

Printed and bound in Great Britain by
Hartnolls Limited, Bodmin, Cornwall

Contents

Preface

Fault finding is a peculiar art that is not easy to categorize. I feel the reason for the popularity of the term 'trouble shooting' is something to do with shooting first and asking the questions later (if at all). As Professor C. Northcote Parkinson said:

> The English are content with diagnosis and are surprised by a request for the cure while the French are concerned only with the cure, happy to leave the diagnosis until afterwards (if at all) . . . the English method is unquestionably more scientific.

While this book is written in the spirit of the English method, its origin, and perhaps its appeal, are 'French'.

Fault finding is an awfully effective way to learn electronics, because it gets to you like nothing else can. It forces you to think inventively but act logically; to harbour fantasy but banish prejudice. What do you remember about your electronics or computing course? What really sticks in your mind?

Evidently the way to becoming a good fault finder is to fix a lot of faults, not to read a book about how to fix faults. I calculated that a lot of set mathematical exercises would be good exam practice but not necessarily apropos intuitive fault finding. (If you are already on a course then you will have plenty of text books with exercises.) It seemed to me that the book should concentrate, as entertainingly as possible, on one simple theme, i.e. the 'variety within unity principle'.

Anyway that was the general idea; I hope that you enjoy this book as much as I enjoyed writing it, but before you begin, let me explain the format.

Although this book begins with voltage, current and resistance, it assumes that you are aware of these things and would recognize a resistor or a capacitor and could perhaps read their values.

The first chapter introduces the theme, the potential divider, in a very gradual way that would appear to be completely at odds with the 'advanced treatment' towards the end of the book – you cannot get from 'Ohm's law' to processors in 200 pages – which is true, as far as *absolute* knowledge goes but it can be done *relatively*, for fault finding.

The first chapter, having introduced the theme the potential divider, goes on to introduce source impedance, not theoretically but as an obvious living thing. The third chapter, Capacitance, inductance and impedance, develops the theme frequency-wise, i.e. reactively. The fourth chapter, Diodes and transistors, expands the theme to its limit for a single component. The first part of the book ends with a chapter on Analogue fault finding expanding the theme to its conclusion for the analogue signal path.

The second half of the book, Fault finding, *depends* on the theme for the subtle bus observations to work.

In this relative sense there is no *disparity* between beginning and end. In another sense, the hardest part of the book is the exposition of the theme and not the recapitulation!

Enough! Otherwise you will start to think this is a book of theory: most definitely not, it is the product of 20 years of fault finding, most of it grafting on production lines. But as I said above, experience cannot be gained from a book, but perhaps the distilled essence of it can, hence the mix of 'English and French' methods.

I have deliberately avoided the explanation of any analogue 'systems' for two reasons: (1) this would dilute the essence of the fault finding method and (2) there are so many excellent specialist and general books on the market that whatever I produced would be a pale copy of these anyway.

In contrast to this I have included brief general descriptions of 'digital sub systems' because this area is still relatively new and not nearly so well supported.

Finally, and above all else, I did my very best to be clear and simple, possibly a little too simple in places, but not (I hope) at the expense of accuracy (some 'simple' sentences get rather long because of qualifying insertions – like this one!).

Robin Pain

Part One

Basics

1 Introduction

Microprocessor systems and analogue circuits are entirely different, forcing this book to be split into two parts, but there is a strong relationship between the two from the fault finder's point of view. Faults themselves are, by their nature, analogue, and a digital fault finder must understand analogue fault finding too.

Analogue fault finding

The radio, television and tape recorder, to name some analogue examples, are primarily amplifiers. A small signal goes in at one end and a much larger one comes out at the other. In a television, the aerial receives signal strengths of the order of microvolts and produces sound and video of the order of tens or hundreds of volts.

The method for fault finding such an arrangement is to start in the middle of the signal path and look for some amplified portion of the original signal. If that signal is present, then we must of course move further along the path towards the output. If the signal were not as expected (missing, or wrong level, or distorted) then we must look more toward the input and so on.

This simple procedure relies on previous knowledge of what the signals along the path should look like or enough knowledge about the local circuitry to be able to guess how it should affect the signals passing through it.

For lower frequency/larger energy signals this procedure is easier to apply but becomes progressively more difficult if the frequencies involved are higher and/or the energies are lower.

Having located the place in the signal path where the signal changes unexpectedly, the fault itself must be found. This requires a special understanding, not only of how circuitry works but also of how it does not work.

2 Voltage, current and resistance

Voltage and current

The size of an electric current flowing along a conductor is measured in amperes or amps (A). An electric current will not flow unless it is forced. This force or pressure is measured in volts (V). The amount of current that flows is proportional to the size of the voltage pushing it, that is, if the voltage is doubled, then the current will double or if the voltage is halved then the current will halve.

When talking fault finding, it is common practice to deliberately 'mix up' cause and effect like this: 'the current through this resistor has doubled therefore the voltage must have doubled'. Whenever a current is reduced, it does not 'slow down', its speed remains constant. It is the *amount* of current flowing that is reduced.

Emf and pd

A voltage between two points is also called a potential difference or pd. It is equally correct to say either 'the voltage is ten volts' or 'the pd is ten volts': both terms are used in practice. Electromotive force or emf is a less common term. It is sometimes used by rf engineers to avoid confusion, i.e. 'input an emf of one microvolt' means by convention that the pd at the input should be half a microvolt.

Resistance

Whenever a current flows along a wire, it always meets with some resistance to its flow. This resistance reduces the size of the flow, i.e. reduces the current. The current flowing is inversely proportional to the resistance, that is, if the resistance doubles then the current flowing is halved or if the resistance is halved then the current flowing will double. Resistance is measured in ohms (Ω).

A large part of fault finding involves voltage measurement. Resistance measurement is hardly used except as a last resort or to prove a deduced fault. Current measurement is not used for fault finding although the total current taken is usually monitored or restricted by the bench power supply (see below). If the fault is excessive current demand, the fault finder will use anything but an ammeter to find the fault.

The relationship of voltage, current and resistance is given by Ohm's Law:

$$V/I = R$$
or
$$V/R = I$$
or
$$I \times R = V$$

where V = voltage, R = resistance and I = current

Understanding how voltage, current and resistance relate to each other is essential to fault finding. The fault finder makes a voltage measurement or two, considers the resistances involved, and by Ohm's law can 'see' the size of the current flow.

Combining resistances

Resistances can be connected in *series* (Fig. 2.1).

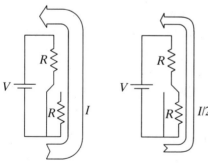

When the extra resistor is switched in, the total resistance increases reducing the current.

Figure 2.1

It is easy to imagine a reduction in the size of the current flow when resistors are connected in series. The rule is: the total resistance equals the sum of the series resistances, i.e.:

$$R_{total} = R_1 + R_2$$

It is a little harder to see what the total of resistances connected in parallel is but consider how the total current is affected in Fig. 2.2.

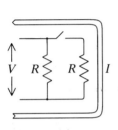 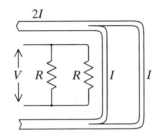

When the extra resistor is switched in then the total current must increase.

Figure 2.2

With both resistors in parallel, the total current flowing must increase. This means that the total resistance must be less, i.e. the total resistance must be less than either resistor on its own.

The total parallel resistance is governed by the 'product over sum' rule:

$$R_{total} = (R_1 \times R_2) / (R_1 + R_2).$$

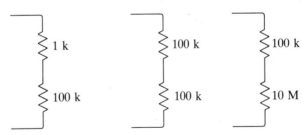

Figure 2.3

What are the total resistances in Fig. 2.3?

Answer: 101 k (just over 100 k), 200 k (double), and 10 M1 (just over 10 M).

Note: This book follows the usual industrial practice of suffixing resistor values with k, M or nothing: e.g. 1 k means 1 kilo ohm = 1 thousand ohms (1 kΩ). 1 M = 1 mega ohm (1 MΩ). 10 = ten ohms (10 Ω).

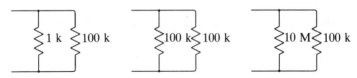

Figure 2.4

What are the total resistances in Fig. 2.4?

Answer: 990 ohms (just under 1 k), 50 k (half of either), and 99 k (just under 100 k).

In other words, for series resistances the larger value dominates and for parallel resistances the smaller value dominates. In both cases, the greatest variation from either resistor happens when both resistors are equal. To put this another way, if a 68 k and a 47 k are in parallel then the 47 k will dominate, i.e. the total resistance must be LESS than 47 k. The smallest value possible is half of 47 k, but that could only happen if the other resistor were also 47 k, i.e. R_{total} must lie somewhere between 47 k and 28 k5. Because 68 k is quite close to 47 k then R_{total} must be closer to 28 k5. By the same reasoning, if a 47 k were in parallel with 1 M, then R_{total} must lie closer to 47 k (than to 28 k5).

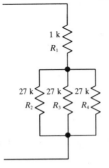

Figure 2.5

What is the total resistance in Fig. 2.5?

Answer: R_{total} = 10 k, i.e. R_2 in parallel with R_3 equal 13 k5, and 13 k5 in parallel with R_4 equal 9 k, i.e. R_2, R_3 and R_4 in parallel equal a 9 k. 9 k in series with R_1 equal 10 k.

Potential divider

Suppose we connect three 1 k resistors in series across 3 voltmeters.

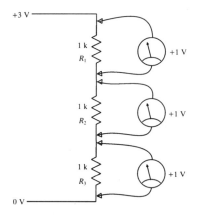

The total resistance is $R_1 + R_2 + R_3 = 3$ k. Therefore the current is $3/3000 = 1$ mA ($V/R = I$). Therefore the voltage across each resistor is 1 mA x 1 k = 1 V. That is, our 3 V supply has been 'spread out' equally across our three resistors (Fig. 2.6).

Figure 2.6

If the supply had been any other voltage then the same would be true, e.g. for a 6 V supply each resistor would have 'dropped' 2 V.

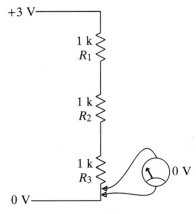

If we call the bottom wire ground (0 V), and measure all the other voltages with respect to ground, then the situation is as shown in Fig. 2.7.

Figure 2.7

9

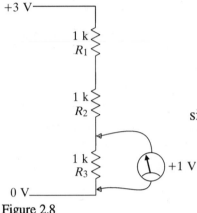

When the R_2/R_3 joint is +1 V the situation shown in Fig. 2.8 prevails.

Figure 2.8

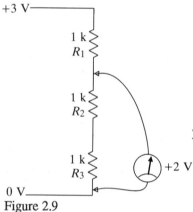

When the R_1/R_2 node is +2 V Fig. 2.9 applies.

Figure 2.9

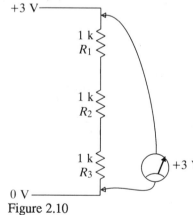

Adding the extra volt across R_1 totals 3 volts. The three equal resistors are dividing the supply voltage into three equal parts (Fig. 2.10). We could make the division continuously variable with a potentiometer.

Figure 2.10

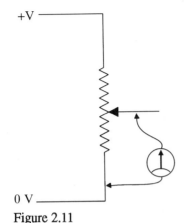

Figure 2.11

With the wiper at the top of its travel we measure the supply voltage V. With the wiper at the bottom of its travel we measure ground. With the wiper anywhere in between we can select any portion of the supply voltage (Fig. 2.11).

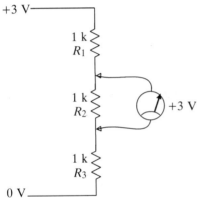

Figure 2.12

Which resistor is open circuit in Fig. 2.12? Before rushing in, consider carefully what would happen for each resistor to be open circuit in turn. Remember that if no current flows through a resistor then there must be no voltage across it, and if there is no voltage across it then the voltage at either end of it *must* be *equal*.

Answer: R_2 is open circuit. If R_1 had been open circuit (o/c) then there would be no feed voltage to the meter and it would indicate zero volts.

If R_3 had been o/c then the same applies, i.e. without an earth return the meter would indicate zero volts.

If R_2 were o/c then clearly the original current of 1 mA could not flow.

So the current through R_1 and R_3 must be zero amps.

Therefore the voltage across them must be zero volts.

If the top of end of R_1 is 3 V and the voltage drop across it is 0 V

11

then the voltage at the other end of R_1 must also be 3 V.

Similarly, if the voltage at the bottom end of R_3 is 0 V and the voltage drop across R_3 is 0 V, then the voltage at the other end of R_3 must also be 0 V.

Therefore the meter has one terminal connected to 0 V and the other end connected to 3 V so it must indicate 3 V.

If you do not understand the last example, then please do not read on, but have a break and then go back and re-read the last section. It is crucial that you understand the last point.

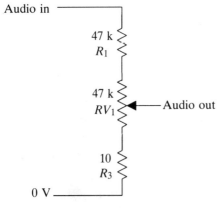

Figure 2.13

Wherever the volume control in Fig. 2.13 is set, whether fully up or fully down or anywhere in between, the output is always at maximum. What could cause this?

Answer: R_3 could be o/c or RV_1 could have a break in its track at the R_3 end. If RV_1 (the volume control) had a break in its track, then the volume would be maximum for all volume settings between maximum and the track break and the volume would be minimum for all volume settings from the break down to the minimum position of the volume control.

Note: For the rather contrived example above, R_3 would appear to be redundant, i.e. it is too small a value to have any influence on the large values connected to it. In cases like this there is nearly always a reason, here it could be mechanical, e.g. R_3 might be a 'wire link' to bridge across some p.c.b. tracking. It is more expensive in production to insert a piece of wire than it is to insert a resistor. Note that special zero ohm resistors are also available just for this purpose.

In the above examples the voltmeter is 'theoetical', i.e. it does not exist except in imagination. A real voltmeter becomes a part of the circuit that it is measuring, e.g. Fig. 2.14.

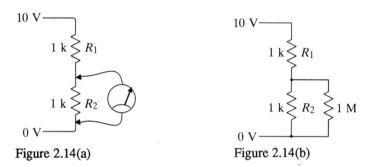

Figure 2.14(a)   Figure 2.14(b)

When measuring the voltage across R_2, the resistance of the meter will change the reading. In the above example, the meter's resistance (1 M say) will appear in parallel with R_2, reducing the value of R_2 very slightly but not enough to compromise the result which will evidently be 5 V. If the resistors were both 1 M instead of 1 k, what would the meter indicate? (See Fig. 2.15.)

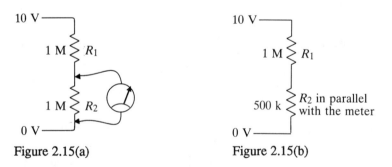

Figure 2.15(a)   Figure 2.15(b)

Answer: 3.33 V, i.e. the meter's resistance in parallel with R_2 will reduce the value of R_2 to 500 k. The 10 V feed will be dropped equally across the total series resistance of 1 M5 therefore 500 k (a third of 1 M5) will drop a third of 10 V. Expessed mathematically this is $R_2/(R_1 + R_2)$.

Resistors do not go short circuit but they do occasionally go open circuit. This tends to happen to the higher value resistors in high energy applications. Lower value resistors survive overloading better. A resistor of 100 Ω for example can be burnt beyond recognition but still be within its tolerance. The reason for mentioning this is that it is always tempting to immediately replace a burnt component, only to find that the replacement starts smoking, i.e. let the original smoke while you (rapidly) find the fault and eliminate it, *then* change the resistor.

Stop light fault example

Good fault finding practice involves imagination, e.g. this fault could be found by measuring everything in the 'standard way'. Can you solve it by thinking alone? A car has a rear light fault. When the vehicle braked only the near side stop light came on and, rather comically, the tail lights on the far side went out. When the brakes were released the far side tail light came back on again (the tail light on the good stop light side remained on throughout as it should do).

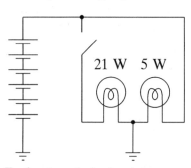

21 W   5 W

Earth returned via chassis

Figure 2.16

Fig. 2.16 shows one side of the vehicle brake/tail light circuit (the other side is identical). When the brake light (21 W) is operated it does not come on but the tail light (5 W) goes out. When the brake light is switched off then the tail light comes on again. Vehicle mechanics will know the answer to this 'stock' fault but may not understand the why of it. Can you figure it out?

Answer: Think in simple logical steps:

1 When the switch is closed the tail light goes out meaning that the *current* passing through it is *less* (not necessarily zero).

2 How could closing the switch do this? One 'obvious' way would be if the filament of the brake light was short circuit (s/c), then closing the switch would s/c the tail light (via the brake light). While this theory would fit in with the brake light failing to come on it is evidently wrong because the vehicle battery would be *shorted out*, fusing the brake light feed.

3 If the brake light was o/c (open circuit) then closing the switch would have no effect on anything, therefore the brake light cannot be o/c.

4 Because the brake light is not o/c, the closing-the-switch effect *must* have something to do with *current*. Not, to be sure, as drastic as the s/c brake light filament above theory, but something similar.

5 This is the 'quantum leap'. In a flash of inspiration, the fault finder guesses the answer.

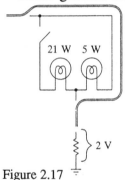

Figure 2.17

6 If the *resistance* of the feed or return wiring has *increased* because of a fault somewhere in it, that would explain all the fault symptoms. Suppose that this extra resistance is enough to drop 2 V when the tail light only is on. That would leave 10 V across the tail light; enough to light it almost fully (Fig. 2.17).

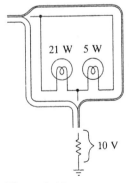

Figure 2.18

7 Now when the stop light is switched on as well, the total current is increased five times (the stop light burns more than four times more power so the current it needs is four times more) so the voltage drop across the fault must increase 5 times to 10 V. This leaves only 2 V (from the 12 V supply) across both the bulbs so the tail light goes out, and the stop light fails to come on (Fig. 2.18).

15

The earth return for both bulbs is returned via a 'mechanical' connection to the vehicle bodywork. Gradually this joint rusts and it goes high resistance.

Observer 'Come on, tell the truth, what really happened?'

Fault finder 'Well we got the vehicle into the garage and started fiddling around with the wiring. The first thing we noticed was that the stop light did come on but very dimly and although the tail light went very dim, it did not go out.'

Obs 'So you never had the s/c-stop-light-theory at all?'

FF 'No. Then we measured the voltages and found the surprising voltage drop across the earthing strap.'

Obs 'Why was it surprising?'

FF 'Because the bolt was very tight and looked ok it did not even look all that rusty! But both lights burst into life at the first touch of the spanner.'

Obs 'So why bother with the what-the-fault-finder-thinks business above?'

FF 'As a memory aid. After finding any fault by the poke and hope method I always do a "postmortem" to see if there was any other, preferably scientific, way to find it.'

Obs 'Ah and this exercise helps you remember the fault details?'

FF 'That's right, next time I see something similar to this I may get a "quantum leap" inspiration because of it! Even if I don't remember anything about the original fault or even pondering about it!'

Potential difference

A voltage, also called a potential difference or pd, is a force existing between two places, e.g. between the plates of a battery. The 'direction' of the force is relative, as illustrated in Fig. 2.19.

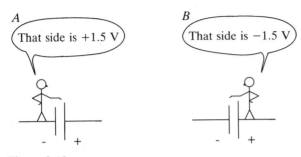

Figure 2.19

Man *A* thinks that he is at ground potential 0 V and so he considers the other side of the battery to be at +1.5 V but man *B* thinks that he is at ground and therefore considers the other side to be −1.5 V

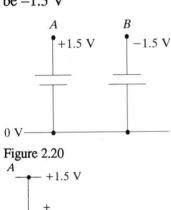

Figure 2.20

Joining the batteries and calling the middle connection ground gives what we call a 'split supply', −V, ground and +V (Fig. 2.20). Note that here the voltage between *A* and *B* is 3 V. This is badly drawn, a better representation would be as given in Figure 2.21.

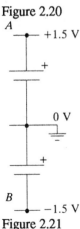

Figure 2.21

It is usual to label diagrams this way. Ground is assumed to be zero volts and all voltages are measured from it, e.g. *B* is −1.5 V and *A* is +1.5 V meaning *B* is −1.5 V wrt (with respect to) ground and *A* is +1.5 V wrt ground.

17

Source resistance

When measuring voltages (or anything else), the meter will always have an effect on the circuitry, it will always place a load on it to a greater or lesser extent and the voltage will fall slightly. The same thing also happens to the power supply, that is, the circuitry connected to it must impose a load on it that pulls the voltage downwards slightly. An example is shown in Fig. 2.22. As the resistance across the dry cell is decreased the current drawn from it must increase.

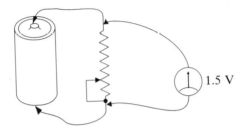

Figure 2.22

If the resistance were decreased to zero then the current would become infinite, which would be true if the voltage remained constant at 1.5 V but clearly it cannot. The dry cell cannot deliver more than a few amps, as this maximum is approached so the voltage of the cell falls off to nearly zero volts as the cell is nearly shorted out (Fig. 2.23).

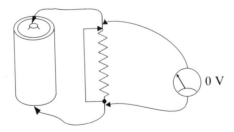

Figure 2.23

The reason for the voltage drop is an accumulation of waste products inside the cell that cannot be removed quickly enough. This acts as a resistance in series with the battery terminals and prevents a large current from flowing (Fig. 2.24).

This internal resistance is called the *source resistance*. Note that it is variable, if the battery is heavily loaded then the source

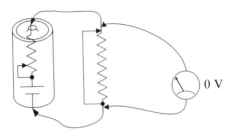

Figure 2.24

resistance will be larger than if it is lightly loaded. (A battery will deliver less energy in total if heavily loaded rather than lightly during its lifetime due to the increased losses in the increased internal resistance.)

All power supplies have internal resistance. This concept will seem odd at first but it simply means that all sources of power are limited. This limitation is a series resistance, it does not matter whether it is chemical, as in batteries, or a combination of things, as in the mains supply (wire–substation efficiency; wire–main generator efficiency; limited maximum water flow to turbines), it can be modelled as a resistance in series with the original supply voltage.

The concept of source resistance and load resistance is absolutely crucial to the understanding of electronics. Hence the emphasis; be sure that you follow everything, do not rush on prematurely.

Current measurement

To measure current the ammeter must be inserted in series with the current flow (Fig. 2.25).

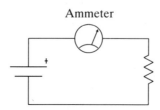

Figure 2.25   An ammeter connected in series.

Evidently the ammeter itself should not restrict the flow too much, otherwise it will spoil its own reading. Therefore ammeters have a low resistance, the lower the better. This, you will recall, is the opposite of a voltage measuring device.

The loading effect of the test gear is not always undesirable. For example, voltage measurement is a poor indicator of the state of a dry cell, without a reasonable load both good and bad batteries might indicate a similar voltage, but, when momentarily shorted with an ammeter, the good cell will supply amps while the nearly exhausted cell will only deliver milliamps.

Fault finder's bench

As indicated above, fault finding is mainly concerned with voltage measurement with the dual trace scope and the DMM. A typical fault finder's bench would look like Fig. 2.26.

The DMM negative lead may stay permanently plugged into the ground reference, depending on the nature of the work, for convenience. Also the flying earth lead of the scope's probe may be omitted, depending on the nature of the work, because the scope will almost always be returned to the chassis of the work via another piece of test gear. That is, the earth of the scope returns via the

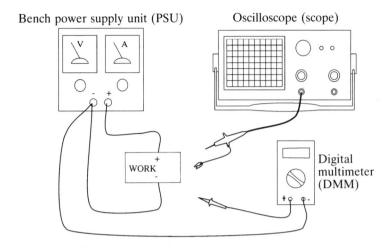

Figure 2.26

mains earth to a signal generator (say) via the generator's input lead to the work's chassis. The grounding of the scope's earth can be a nuisance, the probe's flying earth wire will ground any part of the work it touches!

A nice feature of most bench PSUs (power supply units) is an adjustable current limit. This sets the maximum current allowed providing an element of protection to the work but it can make life confusing if it is set too close to the point where its momentary tripping goes unnoticed.

3 Capacitance, inductance and impedance

Capacitance

A capacitor stores electricity. It consists of a pair of plates set very close to each other but separated by an insulating layer, the dielectric. If a voltage is applied across a capacitor then electrons are driven on to the negative plate while an identical number are drawn off the positive plate. Capacitance is measured in farads (F).

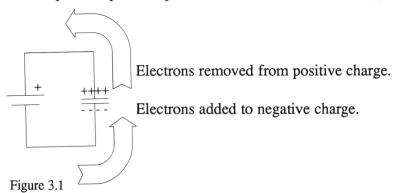

Electrons removed from positive charge.

Electrons added to negative charge.

Figure 3.1

If the battery is now removed, the charge will remain trapped in the capacitor, the excessive electrons on the negative plate will be pressed up against the dielectric desperately trying to return to their homes on the positive plate.

If the plates are short circuited then all the displaced electrons will rush back to the positive plate, the capacitor will become discharged and the voltage across it will return to zero.

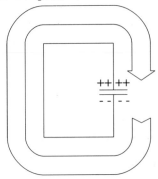

Figure 3.2 Electrons rush back to be reunited with protons.

The charge (or discharge) rate of a capacitor depends on the current flow which is limited by series resistance.

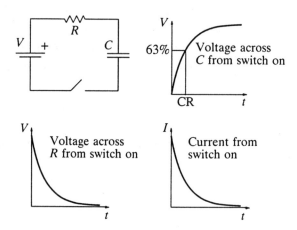

Figure 3.3

With the switch open and the capacitor fully discharged, the voltage across the capacitor will be zero. When the switch is closed, the capacitor begins to charge up. The initial current flowing will be large because the voltage across the resistor is large. With a large charging current, the voltage across the capacitor will start to rise rapidly. As the voltage across the capacitor rises, the charging current must decrease, because the voltage across the resistor is decreasing and therefore the charging current is decreasing. The capacitor will never reach the charging voltage; the closer it gets to it the more the charging current is decreased.

Multiplying the capacitance by the resistance gives the 'time constant' (or CR), the time taken for the capacitor voltage to reach 63 per cent of the charging voltage. That is, if a capacitor of 1 MFD was charged through the resistor of 1 M then the voltage rising voltage across the capacitor would reach 63 per cent of the full charging voltage in one second (CR = 1 E-6 x 1 E6 = 1).

Note two things:
1 The capacitor is 'passing a current'. Although no electrons pass through the dielectric, for every electron that leaves the supply to arrive at the negative plate, another electron leaves the positive plate to return to the supply. To the battery, the capacitor might appear to be a variable resistor (just after switch on).
2 As the voltage across the resistor falls then, as you would expect, so does the current flowing through it, i.e. they are in phase but the same falling current passing through the capacitor is accompanied by a rising voltage across it, a quite different effect.

What is wrong with the circuit shown in Figure 3.4?

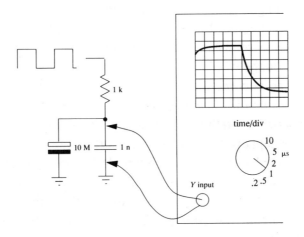

Figure 3.4

Note: This book follows the common practice of suffixing capacitor values with p, n, μ: e.g. 100 p = 100 picofarads, 100 n = 100 nanofarads, 10 μ = 10 microfarads. In industry, capacitor suffixes are often omitted altogether, e.g. 0.01 would imply 0.01 μF, that is 10 nF.

Answer: The 10 microfarad capacitor is o/c. The rising and falling edges of the squarewave are 'rounded off' by the action of the resistor and the capacitance. The time it takes for the voltage to reach 63 per cent (or fall by 63 per cent in this case) is roughly one horizontal division which, according to the timebase setting, is 1 microsecond, which is clearly incorrect. (CR should be in the region of 10 ms (10 M times 1 k). The 'delay' we are seeing is due entirely to the 1 n alone in series with 1 k (1 n times 1 k = 1 microsecond).

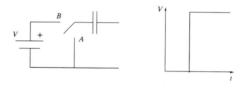

Figure 3.5

In the example shown in Fig. 3.5, if the switch is thrown from *A* to *B* the voltage on the left-hand plate of the capacitor will jump from zero to *V* volts. What will the right hand plate do?

It must also rise instantaneously up to the applied voltage; if it did not then a voltage would appear between the plates meaning that the capacitor had become charged, which, with the right plate disconnected, is not possible. So the changing voltage on the left plate will appear identically on the right plate, i.e. the step waveform will pass through the capacitor as if it were a short circuit.

If a load is placed across the output from the capacitor (Fig. 3.6), then the fast rising edge will still pass through intact but the following steady dc (direct current) level will decay as the capacitor charges up, hence the general rule 'capacitors pass ac (alternating current) but block dc'.

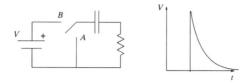

Figure 3.6

If the above battery and switch are replaced with a sinewave generator (Fig. 3.7)*, then the entire waveform will pass through the capacitor.

Figure 3.7

If the input waveform were riding on a dc voltage then that dc component would be blocked by the capacitor (Fig. 3.8).

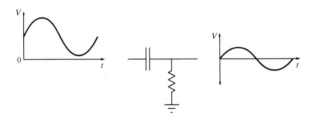

Figure 3.8

The amount of ac passed depends on the value of the capacitor, the resistor and the frequency. The higher the frequency, the more easily it passes through the capacitor. The larger the load resistor, the greater the output will be because the load on it will be less. This arrangement is a potential divider; redrawing it slightly will make it more familiar (Fig. 3.9).

---

* In these and all following diagrams, phase shifts are ignored. A fault finder having to use both scope traces for any reason is rare enough!

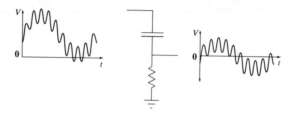

Figure 3.9

The upper resistance is replaced by a capacitor. Note that in Fig. 3.9 the input contains three frequencies:
1 A dc component of frequency 0 Hz.
2 An ac component of frequency f.
3 Another ac component of frequency $10f$.

Also note that each component is reduced by a different amount at the output.

The dc component is attentuated to nothing because, to dc, the capacitor appears as an open circuit (Fig. 3.10).

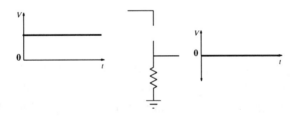

Figure 3.10

The frequency f is attenuated by slightly less than a half (Fig. 3.11),

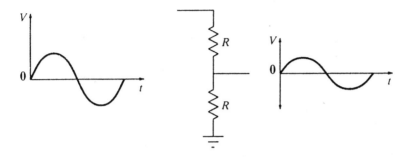

Figure 3.11

i.e. at this particular frequency the capacitor must mimic a resistor of a similar size to R. In the third case, frequency $10f$ is hardly attenuated at all (Fig. 3.12), i.e. at this particular frequency the capacitor appears to be much smaller than R.

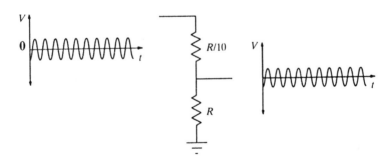

Figure 3.12

29

Capacitors act like a resistor that varies with frequency. This special type of resistance is called *reactance*. When we talk of the resistance of a capacitor we call it *capacitive reactance*. The formula is:

$$X_c = 1/(2 \times \pi \times F \times C)$$

where X_c = capacitive reactance (in ohms);
     F = frequency in hertz;
     C = capacitance in farads.

A resistance like this is called *impedance*. Figure 3.9 is an example of a high pass filter, a filter that allows high frequencies through while blocking low frequencies.

Reversing the divider in Fig. 3.8 produces the opposite filter (Fig. 3.13).

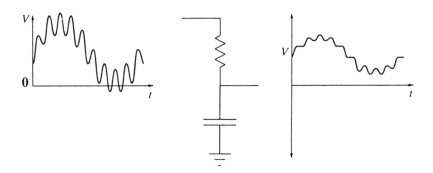

Figure 3.13

In a low pass filter, the dc component is passed without attenuation while the higher frequencies are attenuated more.

What is the fault (if any) in Fig. 3.14?

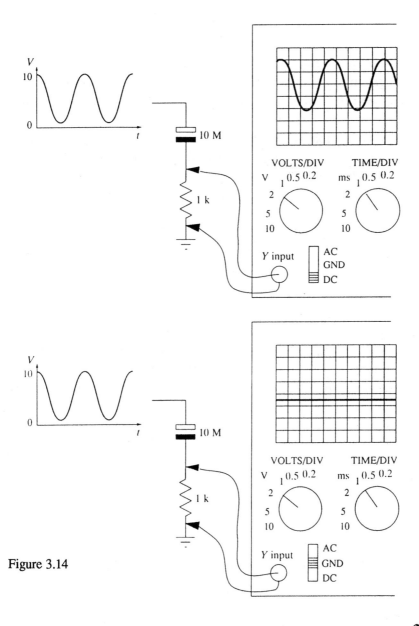

Figure 3.14

The electrolytic is excessively leaky. When the Y input is switched to ground the trace settles just below the centre line of the display. When the Y input is switched to DC coupling, the dc component in the original waveform should be blocked by the capacitor and the scope's trace should be equally disposed about the ground reference, instead the waveform is approximately 3 V above ground. The reason for this can only be that some of the original dc offset must be passing through the electrolytic, i.e. the dielectric is breaking down.

You have probably worked out that this filter should not have attenuated the waveform very much, the reactance of the capacitor at 250 Hz (1 ms per/div = 4 ms/period = 250 Hz) is 63Ω which is nearly 20 times less than the 1 k resistor and therefore should 'dominate it'.

In electrolytic capacitors the dielectric is sustained by the leakage current. If an electrolytic is connected in reverse then the dielectric will be chemically removed and the capacitor will go s/c.

In past times the dielectric of electrolytic capacitors would gradually be absorbed if the device was left standing for long periods without a voltage across it. This was the reason for the periodic switching on once every six months of stored wireless sets or stereograms. I have not noticed any references to 'shelf-life' in modern electrolytics and I assume that nowadays capacitors are more reliable.

Inductance

Whenever a current flows, it creates a magnetic field. This field grows outwards from the centre of the conductor as the current increases or collapses back into the centre of the conductor if the current disappears (Fig. 3.15).

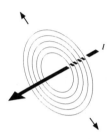

Figure 3.15(a) When a current starts to flow, a magnetic field is created which grows outwards from the centre of the conductor.

Figure 3.15(b) When a current decreases then the surrounding magnetic field starts to collapse back into the conductor.

If another conductor is positioned close by, in the same orientation, so that the moving magnetic field cuts across it, then a current will be generated in it. This new current always flows in the *opposite* direction to the original current (Fig. 3.16).

33

If a current is induced in a nearby conductor then a similar current should be induced in the original conductor, i.e. if an increasing current flows in a wire, then an increasing magnetic field should emerge from it and this moving field must induce a current in that wire.

Figure 3.16 The original current I induces a new current i in a nearby conductor, but only when I is changing. The induced current I always opposes I.

This effect, 'self-inductance', might seem odd at first, but it would be odd if it did not occur. Either *all* conductors in the moving field would experience an induced current or none of them would. It would be inconsistent if the wire carrying the generating current were immune.

By this effect, an increasing current will be reduced by the opposition of the self-induced current, the force driving this opposing current is called 'back emf'. It will be barely noticeable in a single wire (at low frequencies) because the size of the back emf depends on the inductance of the conductor and a single wire has little. At higher frequencies self-inductance is one factor that makes even a single piece of wire a difficult thing to pass a signal down.

Coiling greatly increases inductance as does a magnetic surround, which tends to trap and concentrate the radiating magnetic field. These constructions are called *inductors*. The unit of inductance is the henry. Low frequency (up to audio) inductors have iron cores while higher frequency (from 15 kHz) have ferrite cores, a hard black dense non-conductive brittle material, or no core (air

34

cored). An inductor at UHF (ultra high frequency – 400 MHz) could look like Fig. 3.17.

Figure 3.17 UHF inductor with tap (shown actual size).

As you may imagine, the pcb of layout at these frequencies becomes a critically reactive component in itself.

The inductive reaction to changing currents appears as a resistance, a resistance that varies with frequency and is called *inductive reactance*:

$$X_L = 2\pi \times F \times L$$

where X_L = inductive reactance;
F = frequency;
L = inductance (in henries)

It is exactly the opposite of capacitive reactance. Compare this with the equation above. Inductive reactance is proportional to frequency, capacitive reactance is inversely proportional to frequency. Inductive reactance is proportional to inductance, capacitive reactance is inversely proportional to capacitance. The high pass filter above could be made instead with an inductor as in Fig. 3.18.

The capacitive version of circuits like the one in Fig. 3.18 is more usual, especially at lower frequencies because the equivalent inductors are physically bulkier, heavier and more expensive.

The inductive equivalent of a charged capacitor is rather odd. A charged capacitor is a 'voltage source', i.e. it tries to maintain the voltage between its plates with vigour. If it is short circuited, then it will deliver an enormous current (for a very short time) in trying to maintain that voltage.

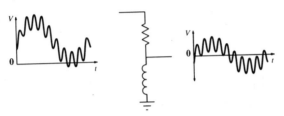

Figure 3.18

An inductor is 'fully charged' when the driving current is constant and the magnetic field is stationary. The 'charged' inductor is a 'current source', i.e. it tries to maintain the current flow through its coil with vigour. If the original voltage across the inductor is now removed the inductor will attempt to keep that original current flowing by increasing the voltage enormously (for a very short time) in trying to maintain that current. The result of this is a high voltage arcing across the switch that interrupted the current flow.

To explain this another way: inductors resist any *change* to current flowing through them, whether the current *increases* or *decreases*. When a current increases, the back emf produces an opposing current; when the current decreases, the magnetic field is decreasing (moving in the opposite direction) and the induced current in now reversed, i.e. flowing in the same direction as the driving current, i.e. adding to it to prevent it reducing – to oppose the change.

A capacitor holds its charge when disconnected (from the voltage): the analogous condition of a 'charged' inductor would be a short circuit. If the inductor was shorted first before the driving current was removed, then the inductor would be able to maintain the holding current itself. Of course the field would vanish fairly quickly because the current would soon be dissipated as heat in the resistance of the windings.

Impedance

The combination of reactance (inductive and/or capacitive) and resistance is called *impedance* (symbol Z).

In practice nothing is a pure reactance or a pure resistance. A piece of wire will have a certain inductive reactance, a certain capacitive reactance and a certain resistance hence the preference for the general term 'impedance' which covers everything.

The true meaning of impedance cannot be had without mathematics because the currents in reactances are not in phase with the voltages: two voltages at different phases cannot be simply added together, it must be done with vectors.

Transformers

If two coils are wound close together so that they are 'magnetically coupled', then any changing currents in one coil will induce changing currents in the other. If a changing voltage is fed across the input coil (the primary) then a similar changing voltage will appear across the output coil (the secondary) (Fig. 3.19).

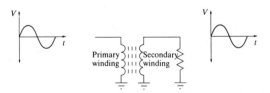

Figure 3.19

If the primary and secondary windings have the same number of turns then the output voltage will be the same as the input voltage. If the load across the secondary is removed, the signal generator feeding the primary will 'feel' nothing connected to it. This is because the self-inductance of the primary winding will be doing its best to prevent any change in current through it, i.e. as the signal generator output increases from zero volts to maximum, the primary will refuse to take any current from it. As the generator swings the output back to zero again the primary will still refuse to accept any current from it, so through each cycle the generator will 'think' that it is not connected to anything because it cannot 'feel' any load across its output.

If the load is now connected across the secondary the whole situation changes. The generator will now 'feel' the load as if it were connected directly across its output. When the generator swings its output from zero to maximum then the primary will start to build up a magnetic field, but some of the field is now absorbed in the load. Without the load connected, the secondary will allow the changing field to envelop it freely without the sightest interference. Of course each time this happens it produces a changing voltage across its output but without a load connected, no current can flow through it. With the load connected, current can flow with the result that as soon as the field starts to build up then some energy in it is absorbed immediately by the secondary to feed the load. The back emf effect is now reduced because the field is reduced and the primary no longer acts as a 'stand alone' inductance but appears to the signal generator to have resistance, which, in this particular case where the primary and secondary turns ratio is 1:1, is the same value as the load (almost, there are slight losses in the transformer).

Transformers are so called because the output voltage depends on the turns ratio between the primary and secondary: V_{in}/V_{out} = primary turns/secondary turns (Fig. 3.20). Here the ratio is 2:1: it can be used either way round either to double the input voltage (step up transformer) or halve it (step down transformer). If the voltage is doubled then the current is halved and vice versa, i.e. the power going in is always equal to the output power (power = voltage × current).

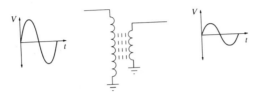

Figure 3.20

Is there anything wrong in Fig. 3.21? This output is clipped, and there may or may not be something wrong in this. A transformer is limited in its capacity to transfer power. The power that a transformer can pass is limited by the size of the magnetic field that it can sustain and this in turn depends on the core size. If excessive power is drawn then the core becomes 'saturated' with magnetic flux before that output need can be satisfied. At this point, the output limits. For economic reasons, a little saturation is generally allowed so the example in Fig. 3.21 may be acceptable or it may be that there is excessive current demand from a fault in the circuitry fed by the transformer.

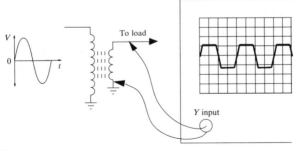

Figure 3.21

39

The short circuit turn

If a single turn, either in the primary or secondary, becomes short circuited, it will act like an independent secondary with a (nearly) zero ohm load. This will absorb any magnetic activity instantly, rendering the transformer useless. Any driving source will 'see' the zero ohm load 'reflected' in the primary and will itself be shorted.

Reflected impedance

The load on the secondary will appear in the primary multiplied by the square of the turns ratio, i.e. a 1 k load in the secondary of a 2:1 step down transformer will appear as 4 k to anything attempting to drive the primary. This property is often used to match impedances (more of this later).

Inductors are rarely used nowadays at audio frequency, excepting psu transformers. In the days of valves there was a common output stage arrangement that regularly failed when operated without a loudspeaker. Can you guess why this was?

When connected, the impedance of the speaker would appear in magnified form at the primary which suited the output valve. When disconnected, does the primary apparently become a harmless high impedance? passing any current changes reluctantly? So it seems, but slowly, the usual standing current would form through the primary (class A bias) and gradually the magnetic field would build up without harm. If the amplifier is now switched off (or an input signal applied), the violent reaction of the inductance would damage the output valve. If the loudspeaker had been connected, then the energy stored in the core would have been 'shorted' out by the few ohms of the loudspeaker.

Transformers offer isolation between primary and secondary, i.e. as far as ac is concerned, the two are connected together as if by a piece of wire, but to *dc* they are totally disconnected. In the valve example above, the primary has a high voltage dc component (with

HT

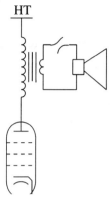

Figure 3.22 When the loudspeaker is left disconnected the output valve would be destroyed.

respect to ground) fed from the HT (high tension) rail but the secondary is 'floating' with respect to ground. Any accidental connection between ground and either end of the secondary would be harmless.

Fuse your loudspeakers!

Modern high power amplifiers are usually 'directly coupled' with no isolation between the output stages and the loudspeaker (LS). Most faults will probably leave the output stage driving the dc supply voltage through the LS destroying the speech coil. To prevent this, fit a fuse between the output and the LS. Choose the value experimentally, so that it blows at near maximum volumes but does not glow at normal listening levels (also known as 'How To Prevent Your Kids Annoying The Neighbours When You Are Out').

If the output stage suddenly swings from its average value of zero volts (from an ac point of view) to a permanent voltage offset (dc fault condition), the fuse will tend to blow before the speech coil burns because the coil is inductive (Fig. 3.23).

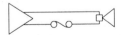

Figure 3.23 Loudspeaker protected with in-line fuse.

41

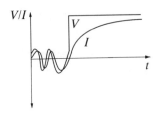

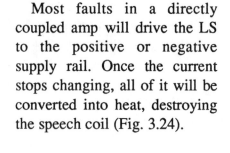

Most faults in a directly coupled amp will drive the LS to the positive or negative supply rail. Once the current stops changing, all of it will be converted into heat, destroying the speech coil (Fig. 3.24).

Figure 3.24

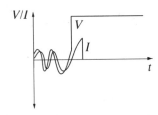

With a suitable in-line fuse fitted a fault will tend to blow the fuse while the current is still changing, i.e. while current in the coil is still being converted into magnetic rather than heat energy (Fig. 3.25).

Figure 3.25

When driven normally, the speech coil behaves inductively with the current flowing through it out of phase with the voltage across it. The voltage rises first and the current reluctantly lags behind as the coil resists the change in the effort establishing a magnetic field. Then the voltage falls and again the current lags behind, reluctant to drop as the coil tries to prevent it falling by converting the magnetic field back into current. In this way little energy is lost in the coil, instead energy is stored in the magnetic field and released back again later. But when the voltage stops moving then eventually the lagging current will 'catch up' and the two will progressively become more and more in phase and less and less energy will be stored in the magnetic field. Instead, the energy will be expended in the resistance of the coil windings. For a 4 Ω loudspeaker driven from +/– 60 V rails this will be 900 watts! With a suitable fuse fitted, the current will be cut off while it is still rising, during the safe period when it is being converted into magnetic flux.

Note. Loudspeakers are 'more resistive' than they are reactive as implied above. Only a small proportion of electrical energy is converted into sound, most of it is wasted as heat.

Resonance

When a (playground) swing is pushed regularly at just the right moments it will swing more and more violently to and fro. This action, between the regular push and the swing's gaining momentum is *resonance*. The swing is in resonance with the pushing. Another example of mechanical resonance is the shattering of a wine glass by singing exactly the right tone to make it vibrate in sympathy. If the singer can maintain the frequency accurately enough, each successive vibration in the glass will be greater than the last until it tears itself apart.

A circuit containing capacitance and inductance behaves in the same way. Before the singer can begin, the wine glass is struck to find its natural frequency, the pitch at which it will resonate. In the example in Fig. 3.26, we 'strike' a resonance circuit with a voltage to make it 'ring'.

If the switch is closed, then the capacitor will charge up to the battery voltage almost immediately. The inductor will initially resist any current flow. If the switch is opened quickly just after the capacitor has charged up but before the inductor can pass any appreciable current, then the following sequence repeats.

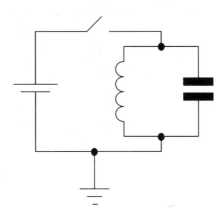

Figure 3.26

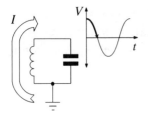

Figure 3.27

1 The capacitor tries to discharge rapidly but the inductor reluctantly passes increasing current so the capacitor voltage gradually falls. Meanwhile, the current increases until it reaches maximum when the capacitor has discharged to zero volts (Fig. 3.27).

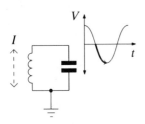

Figure 3.28

2 The inductor, now at 'full throttle' with maximum current flowing through it (maximum field surrounding it), forces current into the capacitor, which greedily accepts it, rapidly charging up in the reverse direction. Eventually all the energy in the magnetic field is expended and the current ceases, leaving the capacitor fully charged (Fig. 3.28).

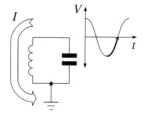

Figure 3.29

3 The capacitor now starts to discharge into the inductor and the current begins to flow again but this time in the opposite direction, slowly at first as the inductor resists it, but by the time the capacitor has discharged, the current has reached a maximum (Fig. 3.29).

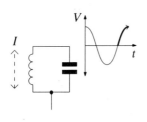

Figure 3.30

4 The inductor is now at 'full throttle' again, with a maximum of energy stored magnetically. It continues the current flow, charging up the capacitor, until all the energy stored in the magnetic field is expended when the current stops and reverses direction, and the entire process repeats from (1) (Fig. 3.30).

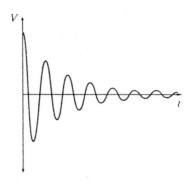

Figure 3.31 The oscillation dies out exponentially but the frequency remains constant. This is the resonant frequency of the LC.

Current will continue to oscillate to and from until all the energy decays away in the resistance of the conductors (Fig. 3.31).

If the switch were closed momentarily at each of the positive peaks, then the oscillations would be maintained indefinitely. This would be equivalent to pushing the swing regularly at just the right moments.

If the switch were closed at one of the negative peaks there would be a clash similar to pushing the swing on its approach. But after a little uncertainty the swing (and the circuit) would be restarted in anti-phase (180 degrees).

The resonant frequency depends on the size of the capacitance and inductance:

$$f = \frac{1}{2 \times \pi \times \sqrt{LC}}$$

where f = resonant frequency;
  L = inductance;
  C = capacitance.

45

All circuits contain some inductance and some capacitance and are prone to resonance. We shall use this 'byproduct' to help isolate faults in microprocessor systems (second section of this book). For the time being, how much information can you get from the example in Fig. 3.32? How well do you know your scope?

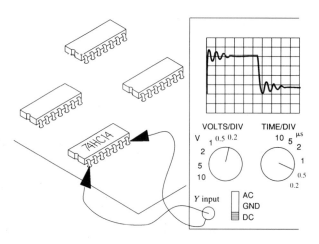

Figure 3.32

The ringing on the trace is caused by the inherent capacitances and inductances of the pcb tracking, the scope's connecting leads, and the components sharing this node. They all combine together to form a tuned circuit that is 'struck' violently by the sudden rising and falling edges and 'ring' as a result.

The ic is a 5 V (maximum) device and yet the scope shows a 6–7 V swing, either the times ten probe has not been properly 'nulled' or it is faulty (the Y gain is set to 0.2 V per division = a reasonable times ten probe setting for 5 V logic signals).

The ringing does seem excessive, it may be that the scope's earth has broken off (usually inside the insulation just at the barrel of the probe) with the result that the earth is now returned via the ring main.

If the probe's earth is unaffected then the frequency of the ringing seems too slow, in which case the time base's times ten magnification may be on. This would make the waveform's period 500 ns = 2 MHz (0.05 µs per division times 10) and the ringing's frequency more plausible.

The scope's probe is on an input (pin 1) which implies a certain distance from an output and therefore the more likely to ring. If the probe had been on an output instead (say pin 2) then the output impedance might have dampened the ringing more. The device has schmitt trigger inputs which implies that maybe this point in the circuitry is lively anyway!

Resonant circuits or tuned circuits are used for selecting or rejecting a narrow range of frequencies (band pass filtering) in one of two arrangements, parallel or series. At higher frequencies, you will recall, an inductor has a higher impedance and a capacitor has a lower impedance and vice versa.

At resonance, these contrary properties combine to produce either an extremely high impedance (parallel aspect) and at the same time an extremely low impedance (serial aspect).

Parallel

Imagine that we are a *voltage generator* about to drive the top of the LC of Fig. 3.33 (with respect to ground) with a sinewave of exactly its resonant frequency. To begin with, the inductor is momentarily high impedance and the capacitor is momentarily low impedance (Fig. 3.33).

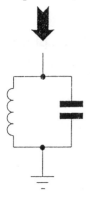

Figure 3.33

For the first few moments we feel a load imposed on us from our having to charge the capacitor: we are imparting energy into the tuned circuit to get it going. After a short time though, the tuned circuit will resonate, copying our own output voltage almost exactly less a tiny fraction. We now feel no load imposed on us because the tuned circuit is moving with us in sympathy. The tiny load we do feel will be due to the trickle of current needed to make good resistive losses.

In other words, from our view point, we are driving a load that takes little current, like a large value resistor, i.e. high impedance.

Series

This time imagine we are a *current generator* inserted in series with the tuned circuit, ready to start driving it with a current of exactly the resonant frequency. To begin with, the inductor is momentarily high impedance and the capacitor is momentarily low impedance (Fig. 3.34).

Figure 3.34

For the first few moments we feel no load imposed on us because the series inductor restricts the current and our voltage output would be forced high to maintain the current. After a short time though, the tuned circuit will resonate, copying our own output current less a tiny fraction. We now feel no load imposed on us because the tuned circuit is moving with us in sympathy. The slight load we do feel will be due to the tiny voltage needed to make good resistive losses.

In other words, from our point of view, we are driving a load that requires little voltage, nearly short circuited, i.e. low impedance.

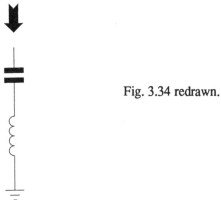

Fig. 3.34 redrawn.

The above explanations might seem difficult, if so, leave them. What they mean is simply this: to an observer, an alternating *voltage* is sustained *across* the inductor and capacitor without any help from the outside world which means *high* impedance; if the parallel impedance were low then the voltage would be 'shorted out'.

Similarly, to an observer, an alternating *current* is sustained in *series* with the capacitor and inductor without any help from the outside world which means *low* impedance; if the series impedance were high, the current would be converted rapidly into heat and lost.

If the resonant frequency is progressively altered, then either the inductance or the capacitance will become dominant, either shorting the signal to ground (parallel) or resisting the signal (series) as shown in Fig. 3.36.

Q factor

If the components were perfect, then the impedance at resonance would be either infinite or zero but resistive losses limit these extremes. Also, to use these devices we cannot avoid loading them externally with our connection.

49

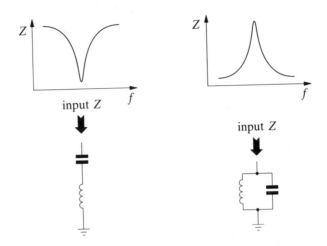

Figure 3.36

To make a band pass filter we need an external resistor (Fig. 3.37). The resistance determines the *quality* or *Q factor* of the tune by 'flattening out' the sharpness of the peak. The same applies to series circuits.

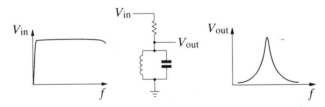

Figure 3.37

The example in Fig. 3.38 shows a notch filter, unwanted frequencies around the resonant frequency are 'notched' out.

50

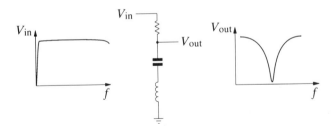

Figure 3.38

Source and load impedance

All energy sources have source impedance, i.e. their outputs are not 'perfect' but limited and this limit acts exactly like a series impedance. The size of it may vary with the load imposed (see the dry cell example in Chapter One) but no voltage source will have an output impedance of zero ohms, otherwise it would be able to maintain its output voltage even across a short circuit delivering an infinite current in the process.

Voltage source

To transfer a voltage with minimum loss, the source (output) impedance should be as low as possible and the input impedance should be as high as possible. For this reason, any source with a low output impedance is called a voltage source. An automotive battery is a better voltage source than the dry cell because it can maintain its voltage better under load.

This may sound confusing. Evidently, the dry cell will only deliver a few amps maximum where the car battery can deliver hundreds of amps (because its output impedance is much less). Which means that a voltage source is a source of unlimited current.

The way to cope with this is to concentrate on the output impedance. If the output impedance is low then evidently it will be able to maintain output voltages better (under load); if it maintains output voltages better then we call it a voltage source.

Current source

To transfer a current with minimum loss, the source (output) impedance should be as high as possible and the input impedance should be as low as possible. For this reason, any source with a high output impedance is called a current source. This too may sound odd because current sources must be able to deliver large voltage swings if they are to maintain a constant current despite the load. If in doubt, concentrate on the output impedance. If the output impedance is very high then a progressive loading will not be able to draw any more current, resulting in the current tending to remain the same.

Impedance matching

When a voltage is transferred almost perfectly (nearly zero ohm output to nearly infinite input impedance) then virtually no current is used which means virtually no power.

If a voltage is transferred and half of it is lost (in the output impedance) then there must be a fair amount of current flowing, which implies a fair amount of power.

If a voltage is transferred and most of it is lost (output impedance high and input impedance low) then the current is a maximum and the voltage is nearly zero, which means virtually no power.

It would seem that to transfer power, the best arrangement might be equal output and input impedances. Which is indeed the case in Fig. 3.39.

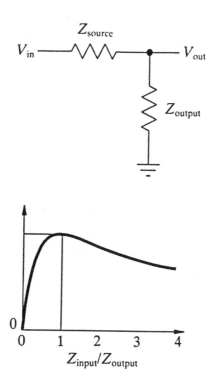

Figure 3.39 Maximum power is only transferred when the source and input impedances are the same.

This with the internal stages of an audio amplifier will probably be the mismatched low source/high input impedance arrangement of voltage amplifiers. (Evidently it makes sense to avoid transferring high power between stages until necessary, so all the gain will be achieved by voltage amplification until the last stage which will convert the amplified voltage from big voltage times small current, to power, i.e. big voltage times big current.)

53

Resonant circuit fault example

The ignition system in Fig. 3.40 gives a 'weak spark', the battery is providing a good 13 V (under load). The spark can only jump about 5 mm and looks 'thin', it used to jump over 10 mm and looked 'fat'. What might be wrong?

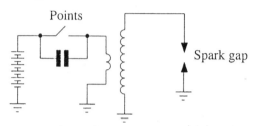

Figure 3.40

What could go wrong? If the coil primary or secondary were open circuit then there would be no spark. If there were one or more short circuit turns in either the primary or secondary then there would be no spark. If the coil were 'flashing over' internally then we should still get a 'fat' spark at 5 mm. The problem must be the capacitor (condenser) open circuit.

If the points were opened faster than an arc could develop between them then all the magnetic energy surrounding both primary and secondary would vent itself across the plug gap. Without the capacitor, as the points open, all the energy would escape in an arc across the points if the secondary spark gap were too large. With a smaller spark gap then some energy would vent via the opening point gap and some via a spark at the spark gap, reducing the energy of the spark.

With the capacitor across the points, when the points are opened a tuned circuit is formed between the primary, the capacitor, and the battery resistance. Provided the capacitor is large enough to effectively 'short' out the opening points, then point arcing is reduced and little energy is wasted across them. The primary and capacitor now resonate (via the batteries' internal resistance) feeding most of the magnetic energy into the spark.

4 Diodes and transistors

Introduction

Up till now we have only talked about passive components. Active components like transistors are much more interesting. The name 'transistor' or 'transfer-resistor' describes its behaviour where a signal passes through, entering a high input impedance and emerging almost unchanged from a low output impedance, a thing not possible for passive components. We call this 'gain', in this instance current gain. Transistors can also provide voltage gain and power gain.

The diode gets its name from the old thermionic valve with two electrodes – 'valve' meaning one way only.

Modern (silicon) transistors and diodes are quite robust and can stand a fair amount of mechanical and thermal abuse but when they fail, they become unserviceable, they do not 'partially' fail, they either work or they do not.

There are cases where a working transistor is replaced with another working transistor to cure a fault but this is a design problem not a component fault! Nevertheless in desperation we will replace working transistors for irrational reasons, 'on the off-chance', but it never makes any difference!

Diodes

The diode has two connections, the anode and the cathode (Fig. 4.1). The cathode is often marked with a 'positive' symbol.

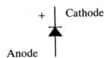

Figure 4.1

A diode will only pass current freely in one directon, and only when the anode is more positive than the cathode by a certain amount.

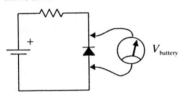

Figure 4.2

When reversed biased (cathode positive wrt anode), the current flow is blocked. And the meter will indicate the battery voltage (Fig. 4.2).

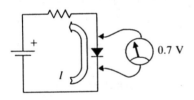

Figure 4.3

When forward biased (anode positive wrt cathode and $V_{battery}$ is more than 0.7 V) then a current flows with a fixed voltage drop across the diode (Fig. 4.3).

The 'forward bias drop' depends on the technology. A silicon diode will be roughly 0.7 V. A schottky (shot-key) barrier diode will drop by roughly 0.3 V.

A diode can be used to turn ac into pulsating dc. This example will only conduct during the positive half cycle (Fig. 4.4). The ac is 'rectified' by the diode to produce pulsating dc.

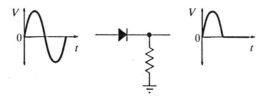

Figure 4.4

Diodes used like this are called rectifiers. If a capacitor is added, it stores charge during the rising edge and releases some of it into the load during the falling edge, 'smoothing' the output (Fig. 4.5).

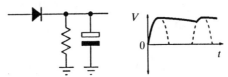

Figure 4.5 The pulses are smoothed by the reservoir capacitor.

Note that the smoothed dc has a characteristic ripple which is still at the same frequency of the original alternating voltage. What appears to be wrong in Fig. 4.6?

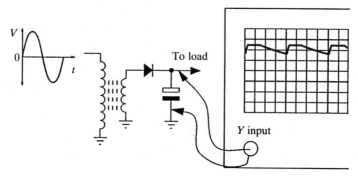

Figure 4.6

The output is clipped and the ripple is pronounced. If the load were faulty it could be drawing too much current. This would excessively discharge the reservoir, deepening the ripple, and it could also saturate the transformer (excessively), worsening any clipping. If the load was within normal limits what could the problem be, especially if the psu was newly built?

The reservoir capacitor might be reversed. It would break down gradually at first, increasing after a few minutes quite rapidly. It would become warm and begin to swell. Some electrolytics have safety vents or weaknesses built into their casings to rupture in a 'controlled manner', but beware! Keep your face away from them!

These half-wave rectifiers use only half the available cycle. Full-wave rectification uses the whole cycle, by reversing one of the half cycles. To help understand its operation, first consider the transformer output in Fig. 4.7.

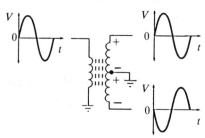

Figure 4.7

Earthing the centre tap of the secondary produces two separate windings. When the tops of both halves go positive together, the bottom half has its top earthed so the free end must go negative.

In other words, the bottom half produces an output that is 180 degrees out of phase with the top half. Arranging diodes across both halves produces full-wave rectification (Fig. 4.8).

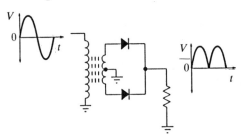

Figure 4.8

During a positive half cycle, only the top half of the secondary conducts. Then, for the negative, only the bottom half conducts, reversing the phase by 180 degrees.

58

All negative cycles are converted to positive by the lower winding and the combined output of both halves provides full-wave rectification. Note that the output frequency is double the input frequency.

Another method uses four diodes in a 'bridge' configuration (Fig. 4.9).

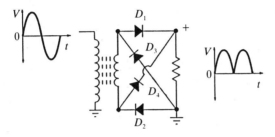

Figure 4.9

During the positive half only D_1 and D_2 conduct (D_3 and D_4 are reverse biased). During the negative half only D_3 and D_4 conduct (D_1 and D_2 are reverse biased). The bridge is usually drawn as shown in Fig. 4.10.

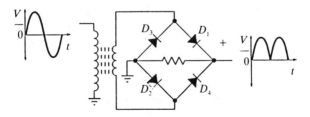

Figure 4.10

59

What is wrong with the output in Fig. 4.11?

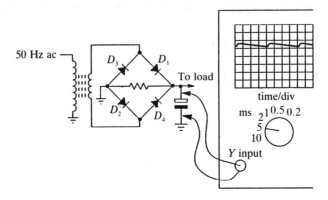

Figure 4.11

The period is 4 divisions and the timebase is set at 5 ms per division; for a 20 ms period this gives a frequency of 50 Hz. The ripple frequency should be twice the input frequency, i.e. 100 Hz, therefore one of the diodes is open circuit.

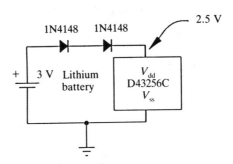

Figure 4.12

The forward bias voltage drop depends on the temperature and current flow. For very small currents it may be less: e.g. the circuit in Fig. 4.12 supplies a backup battery voltage to a CMOS RAM ic.

This device takes so little current at the standby voltage, that both silicon diodes have less than 0.3 V across them. Although the diodes are silicon, the total voltage drop across both of them is less than 0.7 V.

Transistors

There are all sorts of transistors, designed for all sorts of power levels and frequencies, but generally speaking, most of them behave in the same way. There are two types: NPN and PNP (Fig. 4.13).

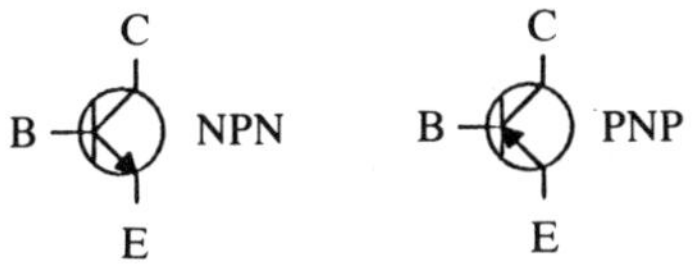

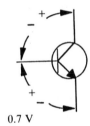

Figure 4.13 An NPN and PNP transistor. The connections are emitter, base and collector.

Figure 4.14 A correctly biased NPN transistor. The base should always be 0.7 V 'above' the emitter, i.e. forward biased and the base collector should always be reverse biased.

By far the most common is the NPN type. A transistor will only work when it is correctly biased, i.e. with the emitter base foward biased and the base–collector reverse biased.

When biased in this way, two currents flow: a small current through the base, I_b and a large current through the collector, I_c. Both flow through the emitter and add together to form a third current, I_e (Fig. 4.15).

61

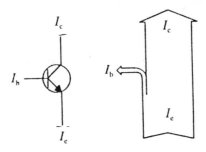

Figure 4.15

If the forward bias across the emitter-base is removed, then all the currents disappear and the transistor is unbiased or switched off.

The large collector and emitter currents are entirely dependent on the tiny base current. If I_b doubles then so will I_c and I_e. If I_b halves then so will I_c and I_e. I_b controls I_c and I_e. For every change in I_b there will be a corresponding change in I_c and I_e. Note that the ratio of base current to collector current is fixed. This is the gain (Fig. 4.16).

The wonderful thing about the transistor is the size of I_b in comparison with I_c and I_e. This is the current 'gain'. Current gains of 100 to 400 are common. A tiny ac current in the base will produce an exact and very much larger copy in the collector and emitter currents, i.e. amplification (Fig. 4.17). If there is a resistor in series with the collector then a corresponding voltage waveform will appear at the collector, i.e. the varying collector current will 'become' a varying collector voltage.

It might seem that the collector and emitter outputs are nearly identical, e.g. with identical resistors fitted in series with each of their voltage outputs nearly identical.

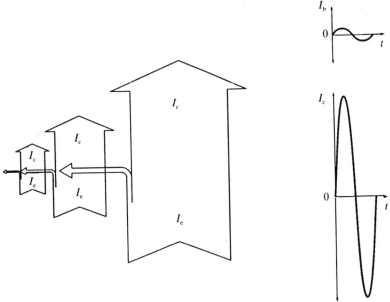

Figure 4.16   Figure 4.17

When emitter and collector resistors are the same then, because I_c is nearly equal to I_e, both voltage outputs are of nearly the same

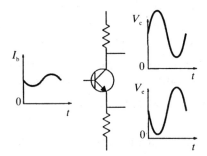

Figure 4.18

amplitude but are of opposite phase (Fig. 4.18). This is deceptive: the nature of the collector and emitter outputs is quite different. The reason lies in the biasing difference between emitter-base and

63

collector-base. This 'glues' the base voltage to the emitter voltage (plus 0.7 V). In the circuit in Fig. 4.18, the base voltage would be identical to the emitter (plus 0.7 V).

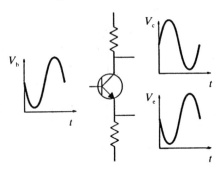

Figure 4.19 The emitter voltage follows the base voltage exactly because the emitter will always be 0.7 V below the base.

If the emitter and base waveforms are now the same, then where is the amplification? The answer is in the current gain. Imagine that we are a negative-going voltage about to 'lean' on the base. At the very first touch, we reduce base current. This produces a massive drop in the emitter current which produces a massive reduction in emitter voltage. The massive reduction in emitter voltage will also 'pull' down the base because the base is 'tied' to the emitter, pulling the base out from under us. The sensation would be like power-assisted steering only more so. If we now tried to 'lift' the base then the increase in I_b would cause a large increase in I_e producing a powerful increase in V_e which would 'push' up the base, again the same effect.

We would conclude that the base was 'as light as air', that it was high impedance. The actual base impedance is the product of the current gain and the emitter resistor.

Emitter–follower

This transistor action (where the emitter copies the base) is exploited in the aptly named emitter–follower, see Fig. 4.20. The output can sustain its voltage because it has plenty of ready current available on demand. From the last chapter you may recall that these qualities are a voltage source and that voltage sources are low impedance.

Hence the name 'transistor', a voltage from a 'weak' source that would otherwise be easily loaded can pass into the high impedance base and emerge 'strong', with plenty of drive power, from the low impedance emitter, i.e. the *trans*fer-resi*stor*, from high impedance to low.

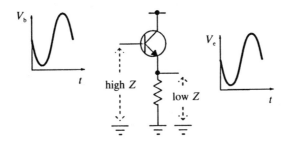

Figure 4.20 The emitter–follower.

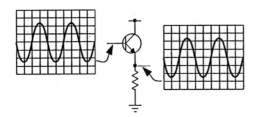

Figure 4.21

What is wrong in Fig. 4.21? Both scope readings have been made with identical settings including dc coupling.

The base appears to be short circuited to emitter because there is no 0.7 V forward bias voltage between them. In a fault like this, the high impedance input, low impedance output of the follower is lost. The (probably) high impedance drive to the base from the preceding circuit is no longer lightly loaded, instead it must drive, through the base/emitter short, and (presumably) low impedance, emitter load,

65

reducing the output of that previous stage, i.e. the waveform in Fig. 4.21 may also be abnormally low.

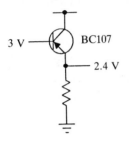

What is wrong in Fig. 4.22?

Answer: The voltages would indicate a forward biased NPN device. The part number confirms this. Therefore the PNP symbol is wrong.

Figure 4.22

What could be wrong in Fig. 4.23?

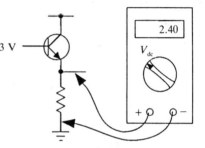

Figure 4.23

Answer: If the emitter resistor was o/c, then the meter's own input impedance would replace it.

It depends on the meter. For a moving coil meter (input resistance of 200 k say) the indicated emitter voltage would remain almost the same. For a DMM (input impedance of 10 M), V_{b-e} would drop to 0.4 V (say) and the meter in Fig. 4.23 would indicate 2.6 V, i.e. there is no fault.

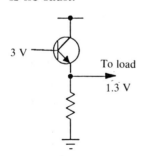

What is wrong in Fig. 4.24?

Answer: The transistor is u/s because the V_{b-e} drop is more than 0.8 V.

Figure 4.24

There is nothing significant in the '1.3 V' level, any V_e between 0 V and 2.1 V would destroy the transistor (if the base remained at 3 V).

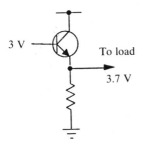

What is wrong in Fig. 4.25?

Answer: The transistor is probably all right. A fault in the load could pull up the voltage at the emitter. Once V_e exceeds 2.4 V the transistor is effectively switched off.

Figure 4.25

A problem like this becomes awkward. Should the load be temporarily disconnected? Or should the transistor be replaced on the off-chance? Cutting pcb tracks is unprofessional (at best) and usually forbidden. One option is to re-check the voltages with the transistor removed.

PNP transistor

The different biasing of b–e and c–b 'polarize' the transistor. NPN transistors will only operate with the emitter voltage less positive (or 'below') both the base and the collector. PNP transistors are the complement of this. A PNP transitor will only operate when the emitter voltage is more positive (or 'above') both the collector and the base. The b–e junction is still forward biased of course and the b–c is still reversed (Fig. 4.26).

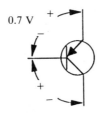

Figure 4.26 This silicon PNP transistor is correctly biased with roughly 0.7 V across e–b and with the c–b reverse biased.

As a memory aid, for the time being, concentrate on NPN types only and forget about PNP. Then when the occasional PNP example crops up, recall the NPN rules and reverse them all.

The prevalence of NPN technology has encouraged the convention of negative earth/positive supply rail. In the early days of transistors, when germanium PNP technology prevailed, the convention was positive ground/negative supply rail.

In the emitter–follower the output is asymmetrical. When the transistor turns fully on (output goes high), a large current can flow through it from ground to the rail. When the transistor turns off (output low) the maximum current flow from ground to the rail is limited by the emitter resistor (Fig. 4.27).

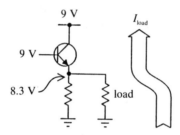

Figure 4.27 When the output is driven high, the maximum available load current is limited by the transistor.

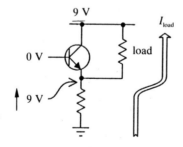

Figure 4.28 When the output is driven low, the maximum available load current is limited by the emitter resistor. As the load demands more and more current, so the output voltage will be 'dragged' upwards by the decreasing impedance of the load.

In other words the follower can 'pull up' much better than it can 'pull down' because the 'on resistance' between emitter and collector is always much lower than the emitter resistor. If this were not so, if the emitter resistor were chosen to equal the c–e on resistance, then when the transistor was switched on, it would not be able to pull up its own emitter load (more than half way), let alone an external one as well. This asymmetry is overcome in the 'push–pull' configuration.

Push–pull

Replacing the emitter load with a complementary transistor provides a huge increase in available current (Figs 4.29 and 4.30). Also, all that current flows through the load, none is wasted in

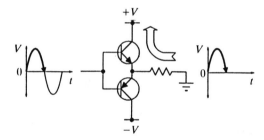

Figure 4.29 On positive half cycles only the NPN transistor conducts.

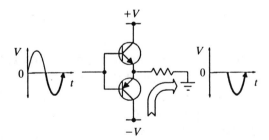

Figure 4.30 On negative half cycles, only the PNP transistor conducts.

69

emitter resistance. When no signal is present, both transistors are off and there is no quiescent current. In the emitter–follower, when no signal is present there will be a heavy quiescent current flowing because V_{out} will have been designed to sit half-way between ground and supply to provide the maximum possible output swing before clipping. Not only is a push-pull output more powerful (per device) but it is also more efficient.

Common emitter

The common emitter is more obviously an amplifier than the follower because voltage is amplified (as well as current). For this reason the common emitter arrangement is called a power amplifier (V x I).

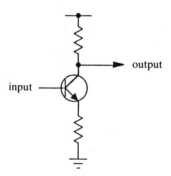

Figure 4.31 The common emitter arrangement.

Suppose we choose equal emitter and collector resistors and deliberately set V_e to a quarter of the supply rail by setting V_b to 2.7 V. If V_b is raised by one volt to 3.7 V then V_e will also rise by one volt to 3 V. What will happen to V_c?

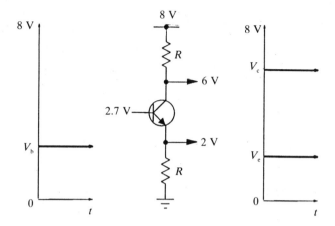

Figure 4.32 Quiescent conditions set up by V_b and the ratio of the collector and emitter resistors.

The answer is that V_c will drop by one volt to 5 V: whatever the increased emitter current is, that extra current flowing in the emitter resistor will appear to us as an extra volt across it. Evidently as ground is fixed at 0 V then the extra volt means that the emitter voltage must rise from 2 V to 3 V. As most of the emitter current also flows through the collector, then that extra current must similarly flow through the collector resistor and so the collector resistor will have an extra volt appear across it (in this example both the resistors are equal). Before this change the collector resistor had two volts across it (8–6), now it must have three volts across it. If the rail is fixed at 8 V then that means that the collector voltage must now fall a volt to 5 V (Fig. 4.33).

If overdriven the collector and emitter outputs 'clip'. At this clipping point the transistor is 'saturated', that is, it is turned fully on, or to put it another way, the resistance between the collector and emitter is minimal. Once the peak positive voltage at the base (that caused the clipped output) starts to reduce, the transistor will drop out of saturation (become less turned on) and the resistance between the collector and emitter will increase and both outputs will emerge from the clipped period to reproduce the driving signal at the base.

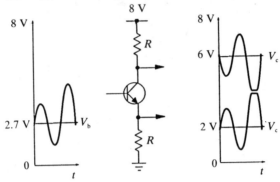

Figure 4.33

Note. There is something very wrong with Fig. 4.33. Can you see what it is?

Answer: Look at the base waveform. When the emitter goes into clipping the base still rises which means that the base–emitter forward voltage must exceed the fixed 0.7 V drop! Which can only happen if the transistor is destroyed. The diagram should look more like Fig. 4.34.

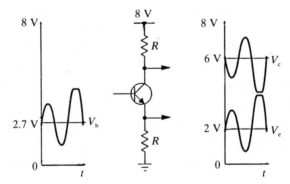

Figure 4.34

If the emitter output limits then the base waveform will also be affected because the base and emitter are glued together by the 0.7 V forward drop. This seems an obvious point to make but in the heat of battle, the fault finder would be inclined to mistake the origin of the clipping. A fault finder who viewed the emitter waveform (or the collector) and then viewed the base waveform would think 'Surely it is coming in at the base and being reproduced faithfully at the output?' (After all, 99 per cent of the time it will be.)

Another point to note in this example is the input impedance of the base. While the emitter is freely copying the base signal then the input impedance will be high (R x current gain), but once the output limits then the wonderful transistor action described above will cease and the input impedance will become R only (because the e–b junction will become a forward biased diode only). This is the reason behind the above possible confusion.

Generally most stages have a buffering effect, i.e. the input impedance is generally high, imposing little load, isolating the input from the output and so preventing any output effects from getting back to the input. But unusually, in the above case, this barrier temporarily drops during clipping as the input impedance is reduced dramatically to virtually nothing (input and output connected together by forward biased diode).

What is the resistance of a forward biased diode?

Answer: Suppose we make a potential divider with a diode and resistor (Fig. 4.35). Evidently, as the resistor is varied from 100 Ω to 10 k, the potential divider will give the same output. What is the resistance of the diode? It rather looks like a 0.7 V battery. The simplest answer is that a forward biased diode is like a s/c with a 0.7 V drop, most peculiar!

To return again to our transistor, what would the effect be if we doubled the collector resistor?

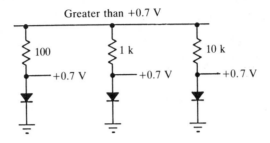

Figure 4.35

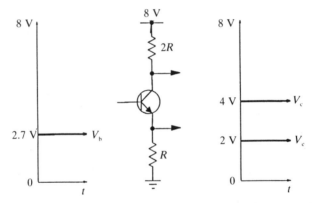

Figure 4.36 Doubling the collector resistor doubles the voltage drop across it.

The way to think about these circuits is like this: start with the base. It has 2.7 V on it. That fixes the emitter voltage at 0.7 V below it at 2 V. Now nearly all the emitter current flows through the collector so the same current flowing through the emitter resistor R also flows through the collector resistor $2R$. Therefore the voltage across the collector resistor must be *twice* as big as the voltage across the emitter resistor to be able to force the same current through it, i.e. 4 V. If four volts are dropped across the collector resistor then the collector voltage must be 4 V below the rail, i.e. 4 V.

74

If the base voltage dropped by a volt (to 1.7 V) then the emitter would drop by a volt (to 1 V). The collector would rise by two volts because the collector resistor, being twice as large as the emitter resistor but passing the same current, must double every voltage change in the emitter.

If a varying signal is fed to the base then the varying output at the collector will be twice the varying output at the emitter (Fig. 4.37).

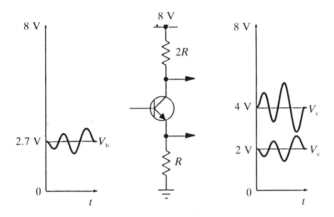

Figure 4.37 The collector output has a voltage gain of two.

Note that the collector has current gain as well. What would happen if the collector resistor were increased still further to $10R$? Well, the collector and emitter resistors form a sort of potential divider with the transistor in the middle. If the top resistor keeps increasing then the voltage at its lower end will keep falling but the collector voltage cannot fall below the emitter voltage. So for values of $3R$ and above, the collector voltage will remain stuck at just above the emitter voltage (the case for $10R$).

What could be wrong in Fig. 4.38? *Hint:* This looks like the $10R$ case above. If the collector resistor were o/c then the collector voltage would fall to ground? Is there another voltage feed to stop it?

Diodes and transistors

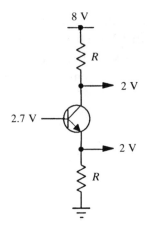

Figure 4.38

Answer: While the collector voltage remains above the base voltage then the collector base junction will always be reversed biased but the collector voltage cannot fall more than 0.7 V below the base: the junction will conduct and, as in the forward biased diode example above, the c–b will become 's/c' together with 0.7 V drop as in Fig. 4.39.

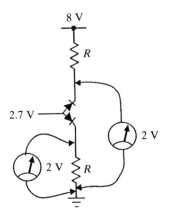

Figure 4.39

If the collector resistor is o/c then the collector voltage will fall to 0.7 V below the base where the b–c will become forward biased because current will be able to flow via the meter to ground.

76

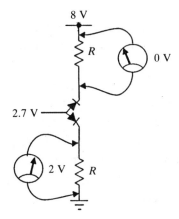

Figure 4.40

If the collector voltage were metered with respect to the rail then the fault would look like Fig. 4.40 perhaps? The c-b 'diode' would be reversed biased so no current would flow through it. Therefore no current would flow through the meter which must indicate zero. Is this true?

No, it would be true for a pair of diodes but not for a transistor. If the collector resistor were o/c then the meter, connected as shown above, would bridge the break with its own resistance restoring the transistor action; restoring the collector–emitter current flow. The chances are that the meter's resistance would be much greater than the emitter resistance with similar results to the above $10R$ case (Fig. 4.41).

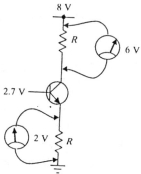

Figure 4.41 Plausible readings.

The voltage gain at the collector depends on the ratio of the collector resistor to the emitter resistor, limited by the quiescent (or dc) conditions (as in the $3R/10R$ case above).

The collector voltage gain signalwise can be greatly increased without affecting the quiescent conditions with an emitter bypass capacitor. The reactance of the capacitor is contrived to be much less than the emitter resistor at signal frequency so the signal 'thinks' that the emitter resistor is much less than it is but the dc levels remain unaffected because to dc the capacitor does not exist. This is also called 'decoupling' the emitter (Fig. 4.42).

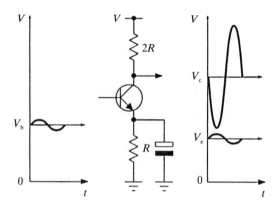

Figure 4.42 Decoupling the emitter increases the collector output.

The voltage gain is dependent on the reactance of the capacitor at the input frequency, the transistor's current gain, the inherent emitter resistance and temperature. Remove the capacitor and the voltage gain is easy to predict: the ratio of collector resistor to emitter resistor. Note that the dc levels (and dc voltage gain) are unaffected by the capacitor and are therefore predictable.

What are the likely faults if the circuit in Fig. 4.42, under identical conditions, looked like Fig. 4.43?

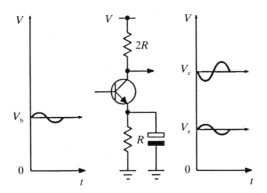

Figure 4.43

Answer: The capacitor is open circuit. The signal voltage at the collector would then drop to twice the swing of the emitter signal (the ratio of R_c to R_e).

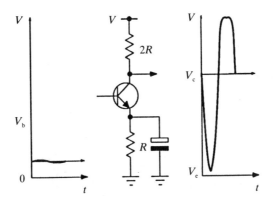

Figure 4.44

What is wrong in Fig. 4.44.

79

Answer: The capacitor is short circuit. This effectively maximizes the collector voltage gain. Note that the base is loading the previous stage in two ways:

1 Evidently to 0.7 V from a dc viewpoint.
2 Less evidently from an ac viewpoint. When the emitter is bypassed then the input impedance looking into the base must become less. If the emitter were perfectly bypassed (or s/c to earth) then the input impedance would be minimum, limited only by the 'intrinsic emitter resistance' of the transistor.

5 Op-amps and negative feedback

Introduction

The (op-amp) is a little self-contained 'system' designed for as much gain and as high an input impedance as possible. Like most systems, any small internal fault will probably produce drastic external symptoms, i.e. op-amps either work perfectly or they do not work at all.

The op-amp cannot 'stand alone', it is always used as a component part of an ingenious feedback arrangement that almost entirely removes the 'variability' of semiconductors. For example, the gain of a transistor could be 100 or it could be 200. Whatever it is, it varies with temperature and collector current.

With an op-amp and a few resistors, it is easy to make an amplifier with a gain of 100, plus or minus 1 per cent (the tolerance of the resistors) using *negative feedback*.

The op-amp has two inputs labelled plus (non-inverting) and minus (inverting) (Fig. 5.1).

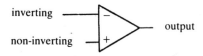

Figure 5.1

Any voltage applied between the inputs will be magnified by the gain of the op-amp before appearing at the output. The gain is enormous, over a hundred thousand (say) and, rather like that of the transistor, is uncertain.

Any signal going into the non-inverting input (wrt the inverting input) will reappear at the output in phase, i.e. not inverted.

Any signal going into the inverting input (wrt the non-inverting input) will reappear at the output in anti-phase, i.e. inverted.

Both these statements are really the same if you think about it, it all depends on your viewpoint (Fig. 5.2). That is, the only thing that matters is the difference between the inputs. If there is 'no difference', i.e. both inputs arc at thc samc voltage then ideally the output should settle at that voltage (in practice there will be a slight error at the inputs which, when magnified by the huge gain, will swing the output to one extreme or the other if we were to tie the inputs together).

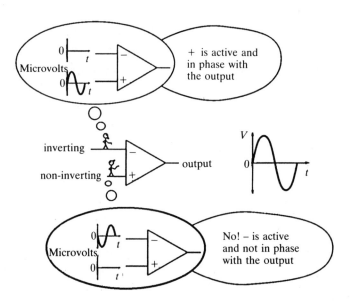

Figure 5.2

From now on, imagine that we are the lower figure standing on the +input and we 'see' one 1 V positive wrt us applied to the – input. If the gain were a million, then the output would try to drive a million volts negative (Fig. 5.3).

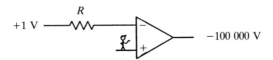

Figure 5.3

Of course, the output would only be able to drive as far as the negative supply rail. An op-amp's output may limit (saturate) a volt or two short of the rails.

If another resistor, equal to the first, is connected between the output and the –input (Fig. 5.4) then the negative output voltage will pull the –input downwards via the new resistor. As the –input falls, the output will rise (in anti-phase) until the *difference* between the –input and the +input vanishes, then the output will hold steady.

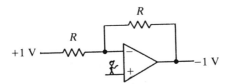

Figure 5.4

To us standing at the +input, this will happen when the output settles at –1 V (wrt us).

To explain this again in another way, imagine a see-saw with its fulcrum at the –input, one end at the input voltage and the other at the output voltage. In addition the length of each arm represents the size of each resistor (Fig. 5.5). If the –input is at zero volts wrt the + input, then there would be no difference and the output would be at zero too.

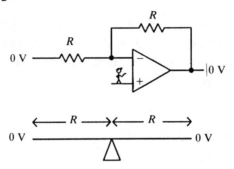

Figure 5.5

The output is low impedance, i.e. the op-amp will be holding the output firmly at zero volts and it would be 'difficult' to move it. If we attempt to lift the input end of the see-saw then it would 'hinge' at the output end, lifting the middle off the fulcrum (Fig. 5.6).

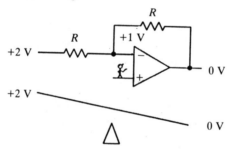

Figure 5.6

Raising the input to +2 V would, it seems, lift the –input to +1 V by the potential divider action (assuming that the output would be held firmly at zero).

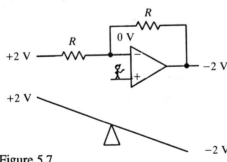

Figure 5.7

But this could not occur: the instant that the –input lifted away from the +input, the output would be driven firmly downward until the –input returned to the same voltage as the +input (Fig. 5.7).

84

Once the output reached –2 V, then the potential divider would have returned the –input back to the + input removing the difference between them. Without a difference the output would hold firm (at –2 V).

If the input voltage was lowered to –2 V then by the same action the output would be forced to rise to +2 V. In other words the output is an inverted copy of the input, i.e. a gain of –1.

If the ratio of the resistors is varied then the gain also varies (Fig. 5.8).

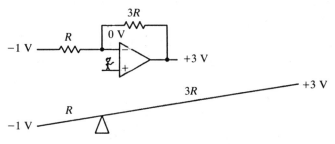

Figure 5.8

Now, the output must drive further to cancel any differences between its inputs. In this case, for a –1 V input the output must drive to +3 V because the resistor ratio is now 3:1.

If the ratio is increased to 5:1 then the gain becomes 5, etc. (Fig. 5.9).

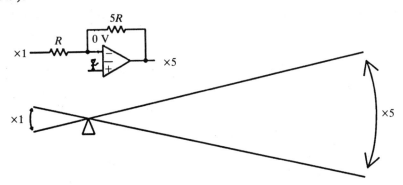

Figure 5.9

Now comes the clever part: the huge gain of the op-amp is now active in *correction*. What was lost in gain has been recovered in accuracy.

If the resistors are equal then the overall gain is 1, the op-amp gain of 1 000 000 is not wasted, instead it is active in making minute corrections to one part in a million, i.e. an accuracy of 0.0001 per cent. If the overall gain were increased to 100, with resistors of R and $100R$, then the accuracy would become 100 times worse, i.e. 0.01 per cent.

Of course the accuracy of the resistors would be the limiting factor and here lies the advantage; we no longer depend on the peculiar nature of the semiconductor but instead we rely on the simpler nature of the resistor.

In the above example (gain of 100) if the op-amp's huge gain were to increase or decrease by a factor of ten, it would not matter because the change in accuracy of 0.01 per cent to 0.001 per cent or 0.1 per cent respectively would be still ten times more accurate, even in the worse case, than 1 per cent tolerance resistors.

The same effect reduces distortion introduced by gain changes within the op-amp, especially as the output swings ever closer to either rail. Without negative feedback the output would be badly distorted should the gain be reduced by a half, for example, as the output nears a supply rail. The above example can stand a gain reduction of ten without unduly affecting the overall gain, i.e. without distortion.

This technique of sacrificing gain for accuracy is called *negative feedback* because the output is 'fed' back in opposition to the input via the 'feedback' resistor.

The op-amp's gain is called the *open loop gain*, i.e. the overall gain if there is no feedback loop. The overall gain, set mainly by the resistors, is the *closed loop gain*.

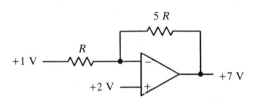

Figure 5.10

Is anything wrong in Fig. 5.10?

No, the input is 1 V negative wrt the +input and therefore the gain of 5 forces the output to 5 V positive wrt the +input.

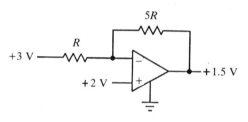

Figure 5.11

What is wrong in Fig. 5.11? The earth connection shown is the negative supply feed for the chip. The output should be driven to –5 V wrt the +input (i.e. to –3 V) but clearly, if the negative supply rail is 0 V then the ic would not be able to drive negative. Many op-amps cannot drive their outputs closer than 1.5 V to their supply rails; this is probably the case here. The +3 V and/ or the +2 V inputs are due to a fault in the prior stages, the op-amp does its best to cope, driving the output as negative as it can.

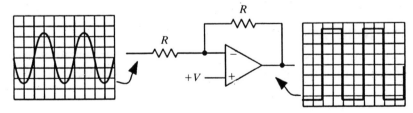

Figure 5.12

87

What is wrong in Fig. 5.12?

The gain appears to be enormous, driving the output to the limit of both supply rails therefore the feedback resistor is open circuit.

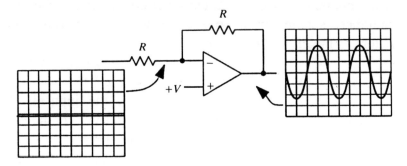

Figure 5.13

Is anything wrong in Fig. 5.13?

No, if the op-amp's gain is 10e5 then the signal level between the +input and the –input will be 10e-5 smaller than the output, i.e. tens of microvolts. The average scope is only sensitive enough to show millivolt levels at best, i.e. the signal must be there because there is an output but it is too small to see.

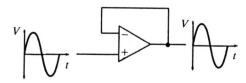

Figure 5.14   The op-amp as a voltage follower.

The follower

If the output is fed back to the –input without the potential divider reduction then the op-amp becomes a voltage follower (Fig. 5.14).

Any voltage at the +input will be accurately tracked by the output, otherwise a voltage difference will start to appear between the op-amp inputs. Note that this arrangement is non-inverting. There is another difference between the follower and the inverting arrangement *viz* input impedance, can you guess what it is?

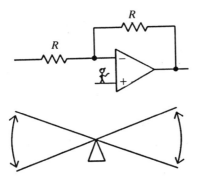

Figure 5.15

The input impedance of the follower is the input impedance of the op-amp, i.e. very high. The input impedance of the inverting arrangement is the value of the input resistor only. This might seem strange, surely the input impedance should be the input resistor plus the input impedance of the op-amp, i.e. even higher? Surprisingly this is not so (Fig. 5.15). As the input voltage varies, the −input appears fixed and immovable, i.e. low impedance. Anything driving the input will 'think' that it is loaded by the input resistor whose far end is held firmly at fixed voltage, the same voltage that happens to be placed on the +input.

The value of this low impedance points depends on the ability of the op-amp to 'defend' it, which is approximately the value of the feedback resistor divided by the open loop gain.

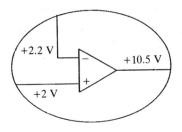

Fig. 5.16 illustrates a common 'beginner's' mistake: because the +input is below the –input, the output should be low, instead it is high, leading to the verdict that the op-amp is faulty. What is wrong with this diagnosis?

Figure 5.16

The inputs are miles apart (in 'op-amp miles') – a catastrophe has occurred and therefore all the rules no longer apply. We cannot predict what the output should be because, for some reason, the output has failed to correct the difference between the inputs. When an op-amp is 'out of control' like this, the output is not guaranteed to be predictable.

It would be fair to say 'Because the inputs are miles apart the op-amp could be faulty' but the first diagnosis is meaningless because once the feedback-to-correct-the-difference-between-the-inputs mechanism has broken down, the inversion/non-inversion rules no longer apply.

Incidentally, although the output may saturate 'correctly' in some cases where the above type of fault applies, in practice, it seems that the output nearly always settles in the 'wrong' state also; the op-amp is almost never the cause!

Frequency response

Frequency response is the big limitation for op-amps; high gain and high input impedance are no problem. For this reason op-amps are common in low (audio) frequency applications but not so at the higher working frequencies.

The problem is to do with 'instability', 'self-oscillation'. As the working frequency increases then combined delays through the op-amp and its surrounding circuitry will progressively delay the output until, at a high enough frequency, the output is no longer inverted but back in phase with the inverting input. The result is the madness of self-oscillation.

To put this another way: the op-amp in Fig. 5.17 will have a certain amount of stray capacitance between its inputs. This means that any changing input signal will be delayed a little before reaching the non-inverting input. This effect 'bends' the see-saw and the faster the input voltage changes, the more the bend.

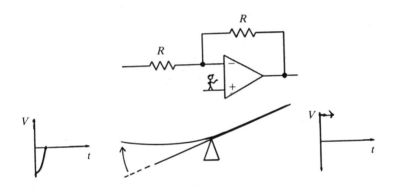

Figure 5.17

Once the changing voltage reaches the −input then a difference will appear between the inputs and the op-amp will begin to amplify that difference, but there will also be a delay in the op-amp itself before this change can affect the output.

The 'input' delay is not important in what follows but it helps to set the scene and start the action.

(a)

1 The −input goes positive wrt the +input but the output cannot counteract because of the delays in the op-amp and the feedback loop so the output is driven to the limit (Fig. 5.18a).

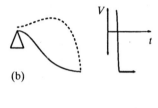

(b)

2 Eventually the delayed output swing feeds back to cancel the difference between the inputs (Fig. 5.18b).

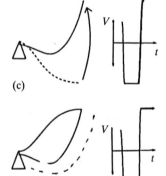

(c)

3 The −input goes negative wrt the +input but, as before, the output cannot cancel the difference and so is driven to the opposite limit (Fig. 5.18c).

(d)

4 Eventually the delayed output feeds back to the −input (Fig. 5.18d).

Figure 5.18

At this particular frequency the op-amp has been made to oscillate. For lower frequencies the 'bending' is less, i.e. at lower frequencies the feedback opposes the input (negative feedback) but as the input frequency increases, the delays cause the phase of the feedback to change progressively until the feedback becomes a half cycle late and actually enhances the input (positive feedback). Note that in the above case, the input is now in phase with the output.

This possibility of self-oscillating is countered in op-amps by limiting the gain at high frequency, either internally or externally or both.

An op-amp can only oscillate if there is gain at the resonant frequency. Most op-amps have their frequency responses tailored to fall off with increasing frequency restricting them to low frequency use only. Some types allow external control of this frequency response via special pins. But even with compensation, extra negative feedback may be required at higher frequencies to prevent instability (Fig. 5.19). At higher frequencies the capacitive reactance 'shorts out' the feedback resistor increasing the negative feedback to prevent instability.

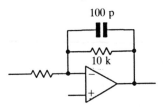

100 p

10 k

Figure 5.19

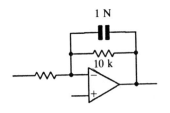

1 N

10 k

Figure 5.20

What is the capacitor for in Figs 5.19 and 5.20? What would we expect if it were missing or o/c? The capacitor will only become significant when its reactance approaches 10 k, i.e. at 16 kHz ($1/2\pi \times F \times X_c$). This lowish frequency (top of the audio range) means that the arrangement is probably for shaping frequency response, rather than for stability. If it were missing or o/c we would expect symptoms to do with bandwidth rather than instability.

Virtual earth

The little figure in the previous diagrams on the non-inverting input is intended to indicate that we are sitting on this terminal and that all voltages are with respect to us, i.e. all voltages are referenced to, hinged about, this point. This departure from the convention of earth as the voltage reference is to avoid the popular term 'virtual earth'.

As explained above, the –input is a low impedance point wrt the +input potential for non-inverting op-amp arrangements. If split supplies are available then the +input is usually taken to earth so that the output may swing equally above and below earth. In this case the –input becomes the so-called 'virtual earth'.

But a lot of equipment manages with a single supply or several supplies all positive wrt a common earth. In these cases, the op-amp input reference is commonly raised to a voltage half-way between supply and earth. By the above convention we should presumably call the –input a 'virtual half supply rail'.

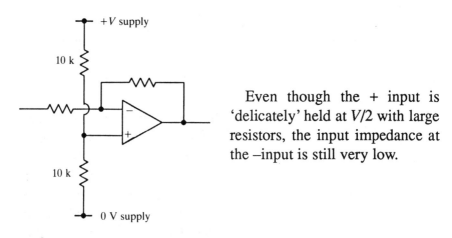

Even though the + input is 'delicately' held at $V/2$ with large resistors, the input impedance at the –input is still very low.

Figure 5.21 The input reference set midway between the supply rails.

If the feedback resistor were 10 k and the op-amp's open loop gain were 200 000 then the impedance at the –input would be approximately 10 000/20 000 Ω = 0.05 Ω wrt $V/2$.

The faulty circuitry in Fig. 5.22 has just come off the production line. What, at a glance, would you suspect first?

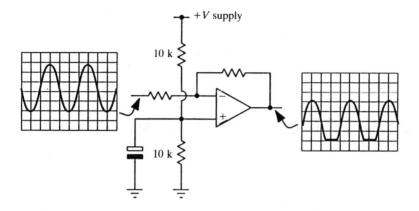

Figure 5.22

If the capacitor were reversed then it would start to leak, more and more rapidly, progressively shunting the lower 10 k, progressively lowering the half rail reference, prematurely clipping the negative half cyles.

Input offset

If, in the above case (Fig. 5.22), the input voltage were set at exactly $V/2$ then ideally there would be no difference between the op-amp's inputs and the output would be forced to settle at exactly $V/2$ (to avoid pulling the $-$ input above or below $V/2$). In practice there will be a little imbalance between the inputs, and the output will be forced to settle slightly 'off centre' to cancel it. The error between inputs is called *input offset*. If we short non-inverting and inverting inputs together, this slight imbalance will be magnified by the open loop gain slamming the output hard against the positive or negative supply.

95

6 Analogue fault finding

The analogue signal path

Generally speaking, analogue circuitry tends to have a single path from the input to the output. The input is usually small and is amplified many times before reaching the output. If the signal level between two stages was wrong, for example too low, then the 'obvious' conclusion is that the gain of the *preceding* stage is too low (Fig. 6.1).

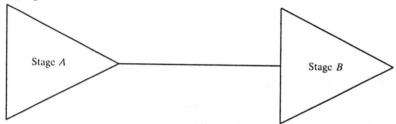

Figure 6.1 If the signal level between stages *A* and *B* is too low it may not be a fault in the gain of stage *A*.

The opening section has emphasized one simple theme, the *potential divider*, the backbone of the fault finding technique of voltage measurement. For every voltage measure we make always consider both the output impedance feeding that point and the small impedance loading it. If the signal between the stages is too small (or missing) then, as well as low gain in the preceding stage, there are two more possibilities: (1) the output impedance of stage *A* is too high; (2) the input impedance of stage *B* is too low.

It does not matter what the stages are, stage *A* might be an entire amplifier and *B* might be another amplifier or stage *A* might be an aerial and *B* might be a feeder or *A* might be one resistor and *B* another resistor, or a capacitor or anything. The output impedance of the first stage will act as the 'top branch' of the potential divider and the input impedance of the second stage will become the 'bottom branch' (Fig. 6.2).

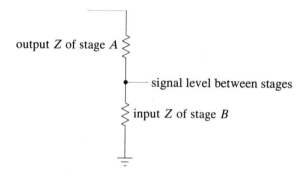

output *Z* of stage *A*

signal level between stages

input *Z* of stage *B*

Figure 6.2

Analogue path fault example

As an example, the circuitry in Fig. 6.3 has a little too much gain. The fault finder is comparing the signal levels with a 'known good working model'.

The output of the first stage is the same as the model, i.e. all the stages up to this point are all right so the fault must be further along. The output of the second stage is the same as the model, however at the junction of the 1 k and 100 Ω resistors the fault symptom appears (the value in brackets is the correct level measured in the model).

Now surprisingly the fault finder goes on to make a further comparison, at the next 'stage', i.e. the junction of the two 1 k resistors. Why would the fault finder do this?

Analogue fault finding

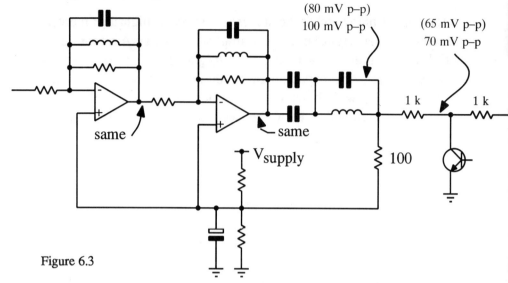

Figure 6.3

Answer: To discover if the input impedance here is greater because if it were greater then that would explain the increased 100 mV level earlier because of reduced loading. To make this clearer: the last 1 k and the transistor and any following circuitry acts as a single impedance, (R_{load}), (Fig. 6.4)

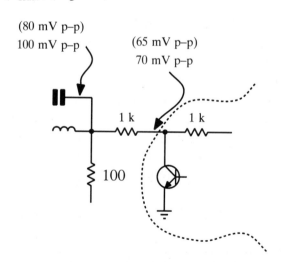

Figure 6.4

98

so that it becomes the 'bottom arm' of a potential divider with the remaining 1 k (Fig. 6.5).

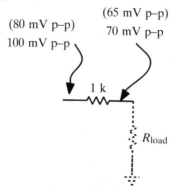

(80 mV p–p)
100 mV p–p

(65 mV p–p)
70 mV p–p

1 k

R_{load}

Figure 6.5

The 100 mV to 70 mV reduction in the faulty unit is the same *ratio* as the 80 mV to 65 mV reduction in the working model which means that R_{load} is the same value for both working and faulty units.

The fault finder has to check this out to be sure that the fault is not due to an increase in the value of R_{load} (imposing less loading effect and explaining the increased levels).

This places the fault before the 1 k.

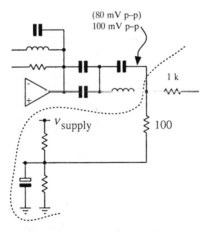

(80 mV p–p)
100 mV p–p

1 k

V_{supply}

100

Figure 6.6 The capacitor/inductor network (driven by the op-amps low Z output) combine together to form the top arm of a potential divider, X_{LC}, (Fig. 6.6).

Analogue fault finding

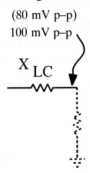

(80 mV p–p)
100 mV p–p

X_{LC}

While the circuitry within the dotted boundary – the 1 k in parallel with the 100 Ω + following circuitry – combine together to form the lower arm (Fig. 6.7).

Figure 6.7

The higher signal level could be due to increased impedance (reduced loading effect) of the lower arm or reduced impedance of the higher arm (more drive).

We can find out which by taking a reading at the lower end of the 100 Ω. What is the fault (Fig. 6.8)?

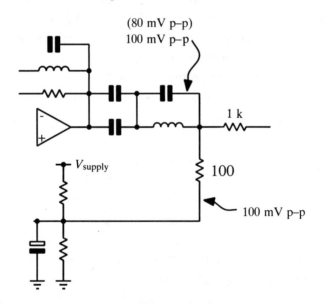

(80 mV p–p)
100 mV p–p

1 k

V_{supply}

100

100 mV p–p

Figure 6.8

Answer: The pcb track between the bottom of the 100 Ω and the 'half rail supply' is open circuit. (If this were not the case, i.e. if the

100

tracking to the half rail supply were good, then that would mean a total lack of decoupling there (the electrolytic u/s or missing) for the 100 mV p–p to exist. In which case the entire circuitry would oscillate violently between rail and ground from the positive feedback.)

Feedback

Negative feedback (and gain) is the stuff of stability, low distortion and predictability. In contrast, positive feedback (and gain) producing instability, ringing and self-oscillation, is usually the result of some fault, unless of course the circuit is supposed to oscillate.

If the output of an amplifier is fed back to its input and the feedback is in phase, then any slight change in the output will be fed back to the input, which will boost the change, etc. until the output is driven hard to one of the supply rails. Then the output will stop moving, the feedback will stop and because the output has reached its limit and cannot increase anymore it must start to fall away from the rail. In this case the feedback will again accelerate the fall until the output hits the other rail then the entire cycle will continue, repeating itself at a certain frequency.

This frequency is special. It is the only one at which the feedback is exactly in phase with the output and it is fixed by the total delay in the feedback loop, i.e. if the total delay is 1 ms then each cycle must equal 1 ms, i.e. a frequency of 1 kHz. Evidently the total gain in the loop must be greater than one, to sustain oscillation, otherwise if the total gain is less than one (i.e. a loss) then the oscillations will progressively die out (ringing).

'Howl around', that piercing squeal sustained by a public address system with the gain set too high, is an example of positive feedback.

Oscillator fault example

In Fig. 6.9 the collector output is inverted and fed back to the base via the two primary windings, i.e. the phase is inverted between base and collector then inverted by the transformer so that the feedback is in phase again, i.e. positive feedback.

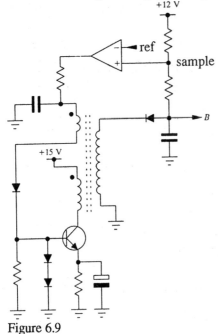

Figure 6.9

There are three examples of feedback going on here:

1 Negative feedback via the op-amp from the output to the input for stability.

2 Positive feedback between collector and base via the two primary windings to sustain oscillation.

3 Negative feedback between emitter and base, suppressed by the emitter capacitor.

When oscillating, the changing currents in the primary windings induce currents in the very much larger secondary winding producing a high voltage output. This ac output is half wave rectified at *B* to a high negative dc voltage. A small sample of this is fed via the non-inverting input of the op-amp back to the base. Clearly this has nothing to do with the oscillating action, it is simply a dc level that varies with the output – if the output should reduce (i.e. if the large negative dc level should become less negative, i.e. more positive) then the dc level fed back to the base would rise.

102

This has the effect of increasing the drive energy of the oscillator to boost the voltage output to restore the ailing secondary output, i.e. this is a negative feedback loop: if the output drops then the feedback, via the op-amp, tries to restore it. If the output should rise then the feedback will reduce the drive energy of the oscillator to prevent that rise. This second negative feedback loop always tries to stabilize the output (under varying loads).

Finally the emitter decoupling, as we already know, reduces ac voltage variations at the emitter which would otherwise 'copy' the base voltage variations 'cancelling out' the effect of the ac drive to the base which would amount to negative feedback.

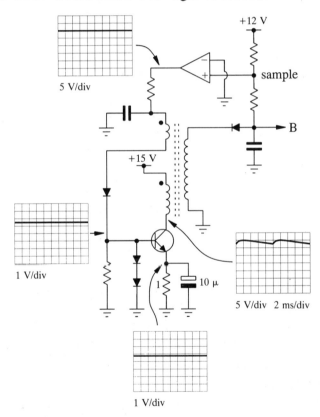

Figure 6.10 For all scope displays, 0 V lies on the centre line.

What is wrong generally, i.e. what are the feedback loops 'trying' to do?

The fault finder and observer examine the waveforms of Fig. 6.10:

Obs 'It's not oscillating, that ripple at the collector is from the psu.'

FF 'How do you know?'

Obs 'Because it has the characteristic 100 Hz frequency of full-wave rectification.'

FF 'What frequency do you expect this thing to oscillate at?'

Obs 'Much higher than that because the transformer has a ferrite core.'

FF 'Why is the 100 Hz ripple so pronounced?'

Obs 'Because there must be a heavy load on it?'

FF 'Correct. How much is due to the oscillator? Look at the emitter resistor.'

Obs 'Ah, it has 0.7 V across it which is quite large considering its only one ohm, that works out to . . . 0.7 A so the poor old psu is struggling . . . 15 V times 0.7 A is 10 W! No wonder I can smell burning!'

FF 'Evidently the transistor is dissipating most of it, what is the purpose of the two series diodes across its base?'

Obs ???

FF 'If they were missing what would happen to the base voltage?'

Obs 'I suppose the +10 V output from the op-amp (via the winding and diode) would try and pull it more positive.'

FF 'And what would that do to the transistor?'

Obs 'It would turn it even more on . . . and it would try and dissipate even more power than it is already.'

FF 'And it would blow up!'

Obs 'So the two diodes in the base are there to protect it, to prevent the base ever rising above 1.4 V!'

FF 'Exactly, which means that they play no part when the oscillator is working normally which implies that the base voltage would normally be less, so the emitter would normally be . . . ?'

Obs 'Less'

FF 'Correct; when it's oscillating, the average emitter voltage is always less. In the old days when we used to get by with only a meter, we could tell if an oscillator was working if the emitter voltage changed when the oscillator was temporarily disabled.'

Obs 'So what's stopping this one?'

FF 'Evidently both primary windings are intact, but one of them, or the secondary, may have a s/c turn which would certainly stop it.'

Obs 'You mean a lack of positive feedback?'

FF 'Yes, or it could be due to a lack of gain. What would cause that, what would reduce the gain of the transistor?'

Obs 'Negative feedback from emitter to base, if . . . the emitter capacitor was o/c!'

FF 'Excellent, so if that was the case then if we temporarily bridge it with another 10 µ then the thing should burst into life.'

It makes no difference.

Obs 'Maybe we should have switched the whole thing off and on to kick the oscillator off?'

FF 'No, I reckon that the ripple in the collector should have been enough, but let's try it anyway!'

It makes no difference.

Obs 'Perhaps the transformer has a s/c turn? Can we check it?'

FF 'Not easily – we'll leave it till last. What about the other negative feedback loop, what is the output of the op-amp trying to tell us?'

Obs 'If the oscillator stops then there will be no secondary output – no large negative dc – so the op-amp will drive its output high to try and increase the drive from the oscillator, which is exactly what it is trying to do, therefore that part of the circuitry is all right or rather not the reason for lack of oscillation.'

FF 'Excellent. There is another possibility, why would a s/c turn stop oscillation?'

Obs 'Because it would act as a separate secondary with a (nearly) zero ohm load on it.'

FF 'AND . . . ?'

Obs 'That would load down the oscillator and stop it; it would not be able to deliver power to satisfy the load and have enough left over to keep itself going.'

FF 'So what should I try next?'

Obs 'Remove the load from the secondary.'

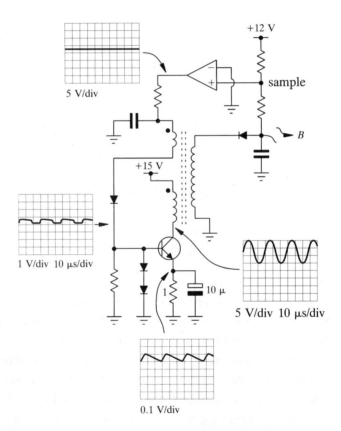

Figure 6.11 For all scope displays, 0 V lies on the centre line.

The fault finder removes the load at B and, given the situation illustrated in Fig. 6.11, when the load at *B* is removed, it works!

Obs 'Well it's certainly going now! Just a minute, the emitter is not following the base, what's wrong?'

FF 'The emitter will only follow the base while the base-emitter is forward biased. In this case the junction is only forward biased at the peaks of the base signal. The transistor only conducts once every cycle for a short time, most of the time it is switched off. Why is this?'

Obs 'Well if it's mainly off then the average current driven through the primary winding will be less.'

FF 'Yes, which means what?'

Obs 'So very little energy is imparted . . . Ah, because we have just removed the load?'

FF 'That's right. Without a load the thing is freewheeling along under its own inertia, it only needs a tiny push once a cycle to keep it going.'

Obs 'And when the proper load is restored the on-time cycle will increase to supply the extra demand for power.'

FF 'Precisely, also the op-amp's output which is low at the moment (to reduce the power) will rise when the load is connected to increase the power to counteract the output drop under load, in the fashion of negative feedback.'

The real fault finding behind the example

A production engineer would snap out this fault in a moment, not just because the correct levels are familiar, but, more importantly to a professional, the physical layout and tracking are familiar.

Anyway you might be interested in the complete mess that I made in diagnosing this fault as an 'amateur'– all professionals are 'amateurs' at each first encounter with a new product (a fact some non-technical managers and customers can never understand).

This fault was in an oscilloscope. The trace brightness would suddenly drop or vanish altogether. Occasionally it would become 'modulated' at 100 Hz, evidently the HT (high tension) section was dropping out of regulation.

These changes were obvious at the oscillator's collector, the amplitude was varying directly with the changes in brightness. Unplugging the secondary loads was the easiest thing to do, one plug to the tube base and another to the 6.3 V heater winding of the mains transformer. This was my first mistake. When either the tube base was disconnected, or the heater connected, then the oscillator recovered its full output without the intermittent fluctuations.

So I reasoned that as it was extremely unlikely that both the tube electrodes and the 6.3 V heater winding would go faulty together, i.e. for both to leak to ground, that the fault must lie in the oscillator's ability to deliver power efficiently – i.e. without a load it seemed to be all right but with a normal load it could not deliver enough energy and so its output declined despite the 'full ahead' reaction of the negative feedback loop.

After fruitlessly changing both the transformer and transistor, which in the case of the transistor was particularly silly, as you could have told me, I was forced to look again to loading at the secondary.

When metered, the tube was free of inter-electrode shorts and the 6.3 V heater winding was entirely isolated from ground or any of the other windings. At the other end of the tube base connector was a plug that allowed easy removal of each of the pin receptacles, so on the assumption that the same sort of inter-electrode short was happening – but only under working conditions – I removed all the pins except the heater pins and switched on. There was no loading effect. I connected the cathode and switched on, naturally, without any grid bias the beam came up very bright and diffused and I hurriedly got out of the way and switched off (for fear of X-ray emission – I have no idea how great a danger this might be but do not intend to find out!). In the brief period that it ran like this, the brightness fluctuated in sympathy with a faint crackle and then became much brighter. This sounded exactly like a tube being boosted (a television engineer's trick to 'rejuvenate' a low emission gun). I reconnected the rest of the pins and the fault had gone. It must have been a partial cathode-to-grid s/c which was inadvertently blasted away with excessive beam current.

Now the reasons or the bad diagnosis were clear. Evidently removing the tube base plug would clear the fault but why would removing the heater supply affect it? It must have been expansion; when the tube base was cold there was no short.

Warning!

The above example is for interest's sake only and not intended to encourage anybody to tackle high voltage equipment, e.g. a colour television can still pack a mighty 25 kV punch long after it has been disconnected from the mains. Even equipment with small CRTs will store voltages in excess of 10 kV.

Directly coupled loops

What do you do if the signal path is a loop and all the interconnections are directly coupled?

High power audio amplifiers, servo amplifiers, and other control circuits may have one or more directly coupled negative feedback loops. This might appear to make fault finding tricky because a fault anywhere in the loop will affect all the voltages around it, as in Fig. 6.11, for example. A fault in the (upper) feedback loop will evidently affect all the voltages around the loop (Fig. 6.13).

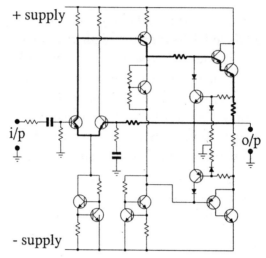

Figure 6.12 During positive going cycles, the highlighted path is the feedback loop. (During negative half cycles, the feedback loop is returned via the lower output pair.)

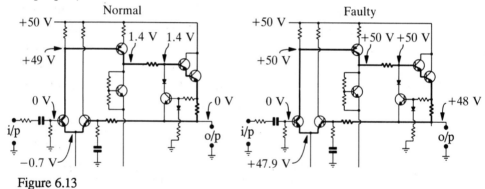

Figure 6.13

Can you find the fault in Fig. 6.13?

Answer: The PNP transistor is s/c between base and collector or between collector and emitter (or s/c between all three).

Begin at any place on the loop, and follow the path checking each stage in turn for correct operation, e.g. let us start at the output (+48.3 V). The emitter of the next stage should be –0.7 V lower than

110

the base so assuming that the 48.6 V level is getting to the base, the emitter should be 47.9 V, which it is, so that stage is functioning properly.

The next stage has 47.9 V on its emitter and 0 V on its base. This does not mean that the transistor is destroyed (if it had 47.9 V on its base and 0 V on its emitter it would!); it means that the transistor e–b is reversed biased, i.e. switched off. If the transistor is switched off then no collector current will flow and if no collector current is flowing through the collector resistor then there must be no voltage drop across it (if it is intact) which means that both ends of the resistor will be at the same voltage, i.e. 50 V. So this stage is working correctly (the voltages around the stage are violently wrong but the stage is operating correctly).

The next stage has 50 V on its base and 50 V on its collector; evidently it cannot be forward biased because the emitter is connected to the 50 V rail (we assume), anyway, where could the necessary 50.7 V come from? So the transistor must be turned off (it is PNP and is therefore 'upside down') which means the collector voltage should be a minimum, i.e. minus 50 V (in the same way the preceding NPN stage was a maximum at plus 50 V), i.e. this stage is faulty but let us continue through the loop.

The next stage has 50 V on its base; both emitters should give a combined drop of 1.4 V, i.e. the output of the second emitter should be 48.6 V which it is, so this stage is all right, i.e. all the stages have been checked as functioning correctly except the PNP stage.

Take five

When I applied for my first 'audio' job I had had three years of rf and a few months of digital experience. I thought the audio job would be fairly easy by comparison. Although I had no audio experience, I really pushed hard in the interview and as a result got the job.

During my first day I found myself trying to repair a high power dc amplifier which a couple of sales representatives needed urgently. As is usual for me, I was very nervous, so much so on this occasion that when one of them asked 'How's it going?' only grunting sounds and mumbles came out of my mouth. I have no idea what they thought, but I knew that after two weeks I would start to understand things and nobody is sacked on the first day, surely. However, they rang for the designer/proprietor to come down to production to repair a 'hard fault'.

He arrived, and after 30 seconds turned to me and said 'Bandwidth?' I grunted in a sort of neutral tone without knowing what he meant. He then started comparing the two channels (this was a stereo amp) visually and spotted a different sized capacitor. It took him two minutes; I considered running away . . .

The thing that kept me going, and this is the point of this story, is that I understood the basics of fault finding which are the difficult things to master. In contrast a system is much easier to understand.

After two weeks I started to relax and after two months I really enjoyed the work. Audio systems turned out to be much more interesting than rf, especially sound effects.

Bandwidth

The method used by that particular audio company to check the bandwidth of their audio product was beautifully simple: could it pass a squarewave cleanly?

The way the method works is based on the theory that a perfect squarewave is composed of an infinite number of (odd) harmonics.

The simplest waveform is the sinewave (Fig. 6.14). Using only sinewaves, of arbitrary frequencies, amplitudes and phases, any waveform can be synthesized. Or, conversely, any waveform, no matter how complex, is only a combination of sinewaves in the last

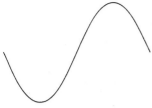

Figure 6.14 The sinewave is 'pure',
i.e. it only contains a single frequency.

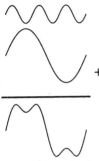

Figure 6.15

analysis (Fig. 6.15). For example, if a sinewave of frequency *f* is added to its 'third harmonic' (same phase) with an amplitude of one third of the original, then a more complex waveform is produced. If this process is continued for all the odd harmonics the waves as illustrated in Fig. 6.16 are produced and when all the odd harmonics have been added the result is a perfect squarewave. A squarewave, although it looks simpler, is in fact infinitely more complex than a pure sinewave!

Fundamental frequency *f*, amplitude *a*

Third harmonic 3*f*, amplitude *a*/3

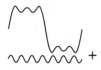

Figure 6.16

Fifth harmonic 5*f*, amplitude *a*/5

113

Figure 6.16 continued

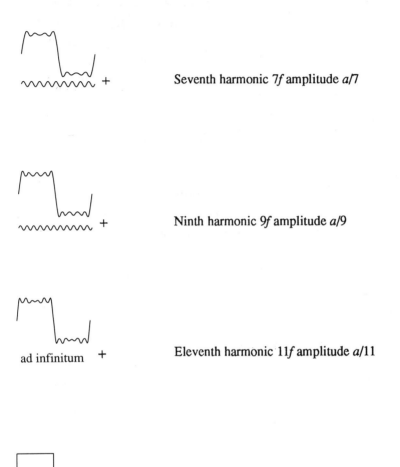

Seventh harmonic 7*f* amplitude *a*/7

Ninth harmonic 9*f* amplitude *a*/9

Eleventh harmonic 11*f* amplitude *a*/11

a squarewave

If the frequency response of a stage is flat, then a squarewave will pass through without distortion. If any part of the bandwidth is cut or boosted then the squarewave will be distorted in a recognizable manner (Fig. 6.17).

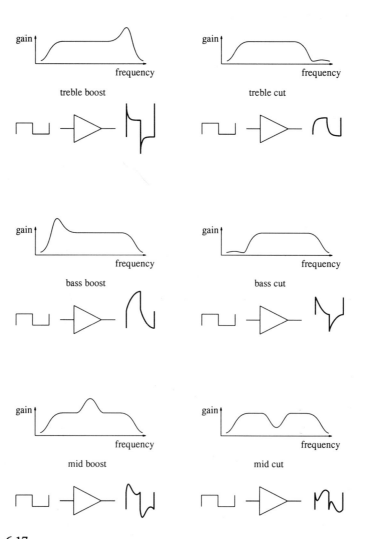

Figure 6.17

115

Is anything wrong in Fig. 6.18?

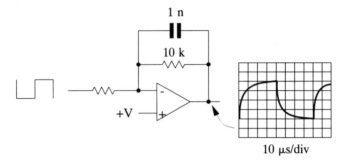

10 μs/div

Figure 6.18

No, the bandwidth looks all right.

At lower frequencies where the reactance of the capacitor is much greater than the 10 k resistor, the 10 k resistor 'dominates' and the total value of the feedback impedance is slightly less than 10 k.

At higher frequencies, the reactance of the capacitor decreases until it equals 10 kΩ at around 16 kHz. Here, the total impedance is 5 k (both values are in parallel). At higher frequencies still, the decreasing reactance of the capacitor is dominant, rolling off the output, cutting the treble response.

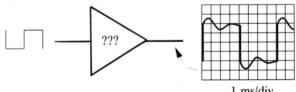

1 ms/div

Figure 6.19

The time constant indicated by the scope is slightly less than one division, which is near enough to the 10 μs time constant of 10 k multiplied by 1 n (with tolerances of scope 5 per cent, resistor 5 per cent, and capacitor 10 per cent).

What is the approximate frequency response of the unknown circuit in Fig. 6.19?

Answer: The 'leading' edges show no sign of treble boost or cut. There appears to be 'mid range' boost (not 'cut') ringing periodically at 3 divisions, i.e. 330 Hz. There is no sign of bass lift or cut.

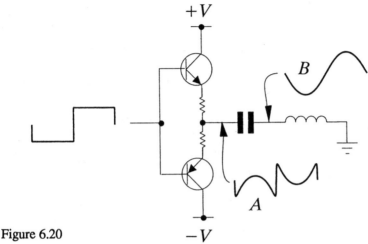

Figure 6.20

Do the waveforms in Fig. 6.20 make sense to you? The series tuned circuit has a low impedance at resonance. The emitters both 'look' into this series low impedance whenever they output the resonant frequency. For all other frequencies, the emitters 'look' into a high impedance (and will be less loaded down).

The squarewave output from the emitters contains many frequencies, all but one are above resonance and will only be lightly loaded. These combined frequencies will appear as the waveform at *A*. The fundamental, however, will be heavily loaded by the divider action and will be almost absent at the emitters, i.e. 'mid range cut'. This frequency will resonate the tuned circuit and will appear as a maximum across the parallel portion of the tuned circuit at *B* which of course is high impedance. Adding the two waveforms back together would restore the original squarewave.

117

Noise

Where there is a lot of gain, there will also be noise. Usually, the noise source will be in the first stages because 'similar' or even noisier sources in later stages are too far down the chain to be amplified to noticeable levels. Even so, occasionally, a later stage can be noisy enough to dominate.

Trying to find a noise fault with a scope is futile. The ear's logarithmic response to sound fools us into believing that different volumes are very much closer together than they really are. The noise that we hear when the volume is full up, which sounds quite loud, might just be visible with the scope cranked up to 5 mV per division, at the output of the PA (power amplifier). As the source of the noise will be in the earlier stages, i.e. hundreds or thousands of times smaller, searching for it with the scope will be futile.

The usual procedure is to wind all level controls to maximum, and listen for any changes in the character or level of the noise whilst 'manipulating' the input stages.

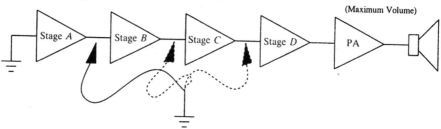

Figure 6.21

One way to do this is to isolate the stage by progressively 'removing' all earlier stages from the signal path (Fig. 6.21). If stage *C* is the culprit, then shorting *A* or *B* to ground will make no impression but when stage *C* is shorted then the noise at the PA will vanish. Evidently the temporary short to ground in Fig. 6.21 is symbolic; a safe way to block the signal path electrically would be as illustrated in Fig. 6.22. To short circuit pins 1 and 2 for the first

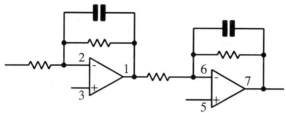

Figure 6.22

stage or pins 6 and 7 for the second stage with something handy like the tip of a scope probe. This would change the value of the feedback resistor to zero ohms; i.e. the gain would become 0 divided by the input resistor value, i.e. zero.

This also applies to quad packs (four op-amps in a 14 pin dil package), the outputs are at the four corners, 1, 7, 8 and 14 and the associated inverting inputs are pins 2, 6, 9 and 13 respectively. (Evidently it is more stable to situate the inverting input closer to its output rather than the non-inverting input.)

The classic way to show up a noisy component is freezer/heat gun (or hair dryer). Spray a large area first and wait a few seconds. If the noise persists spray another area; if the noise suddenly reduces apply a little heat to restore the noise again (and confirm the area) and repeat, gradually localizing the exact component.

Be aware that you may induce 'noise' accidentally via condensation; if the circuitry is high impedance and high gain, it might self-oscillate. It is also very easy to 'excite' the input stages into producing hundreds of watts of output at the PA! A painful experience if you happen to have your head inside the loudspeaker at the time!

Self-oscillation, whatever the cause, is sometimes chaotic and might sound rather like noise. The difference will be obvious on the scope; the noise will probably be too small to see while self-oscillation will be swinging rail to rail: it only sounds quiet because the high frequencies are at the limit of our hearing.

Instability

Self-oscillation or instability is often caused by poor earth returns (Fig. 6.23). Check the input shorting links are 0 Ω when the jack

119

plug is removed. Ensure that any output power modules are screwed tight to the chassis. Look at the surface of the pcb – has any liquid been spilled on it? If so, clean it all off, especially under any ic packages – it may be conductive (or become conductive in humid conditions). Check that any pots and faders are screwed tightly to the chassis, especially input gain pots.

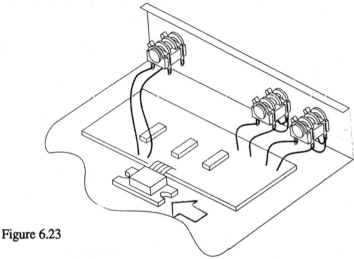

Figure 6.23

Learn to listen

It takes time to learn to recognize particular sounds. I used to dismiss those hi-fi enthusiasts who 'overreacted' to different sound systems, when I could never tell the difference. It seemed to me to be all-in-the-mind. One day, someone commented that the left hand tweeter was down, as indeed it was. This person had noticed the fault instantly and I was impressed, so I started to listen, not just to different loudspeakers but to everything.

The brain's ability to analyse sound is incredible, we can tell in a moment if that half heard conversation in the next room is real or reproduced, no piece of test equipment could do it. Evidently speech is very familiar and it is easy for us to notice deficiencies in its quality which is not the case for unfamiliar sounds, i.e. hearing ability has to be *learned*. Those who drive will know that after years

120

of hearing the same sounds, something as slight as 'pinking' (pre-ignition) is highly intrusive to the driver while the passenger cannot understand what all the fuss is about, let alone make out the sound.

Impossible faults

'Impossible faults' are those frustrating faults that apparently only involve a handful of components which we change and yet the fault remains. In order of likelihood these are my experiences:

1 There is no fault.
2 A design 'feature'.

Possibility (1) is not as strange as it sounds. Sometimes what was assumed to be a fault symptom turns out to be a normal effect in working units. A mistake which is quite easy to make for 'marginal' cases.

There is no fault – example 1

I had successfully tested and aligned nine out of ten units (small low frequency rf receivers) when the last one showed signs of instability. If the receiver was slightly off tuned, it would function acceptably but when properly 'peaked up' it would self-oscillate. It was very difficult to deal with because everything had some influence, if anything was touched near the front end, it would burst into self-oscillation. If a large rf level was blasted through it then the overall gain was found to be identical to the other nine units (that were functioning properly). But of course then the large level would force the AGC to reduce the gain and there would then be no signs of instability.

The Q of the inductors felt exactly the same as the previous nine that had passed, i.e. when they were tuned up, no particular coil was especially 'peaky'; the bad unit felt the same as the good ones. After a tedious session of comparisons between the front ends of both good and bad units by blasting in high levels (pre the AGC) and after changing the front end transistors (swapping them over between good and bad units) we decided that there was no fault and gave up for the day.

The next morning, returning to the job bright and fresh we found the fault in two minutes, not by any scientific approach but by sheer luck. It was due to the underlying conductive anti-static mat!

When the unit was lifted clear of the mat, it became stable, when it was lowered on to it, it became unstable. So why did the previous nine units all work? It was because the component legs are cropped to different lengths. Our 'faulty' board must have had a combination of legs standing proud to form a positive feedback path through the mat.

There is no fault – example 2

I am indebted to a Mr T. Phillips for this story which happened to him when he was field servicing in the early days of ultrasonic remote control television. He was called out to the same address, time and again, for the same problem: 'the television keeps changing its own channels'.

At each visit the television was always fine. But the customer insisted that the remote control was faulty. Some evenings he would have to change channel several times because it kept changing back again of its own accord. Then at other times it would suddenly change channel for no reason.

The mystery was solved accidentally when, during a call, another television engineer turned up outside the house next door. Naturally they got talking and the other engineer said that he had this impossible fault, with an ultrasonic remote control television . . .

As you will have gathered by now, these neighbours had identical sets and remote controls and their televisions were on either side of an adjoining wall.

Design 'feature'

Design feature problems are easy to spot in production, where there will be a regular quantity dropping out with the same symptoms, but they are difficult for the field engineer to identify if one should slip through.

A customer complained that the keyboard of his computer would produce the wrong characters when it was first switched on, but not always.

The fault could never be reproduced in the repair shop, but the customer insisted, adding that it happened particularly badly on one occasion when he took the computer out of his car into his bedsit and immediately switched it on. This was the clue, the fault might have something to do with condensation. It turned out that the impedance of the keyboard tracking was highish, 10 k, and the cross talk induced on the 'non-selected tracks' by the active ones during the scan was rather large. It only took a little conductive condensation to 'push it over the edge'. The manufacturer had already discovered this and had modified the latest build to resistive pull ups of lower value. This modification solved the problem.

RF

Fault finding at rf is less straightforward than at audio frequencies, and the higher the frequency, the more awkward it becomes.

One problem is coupling: how can we connect test equipment? Even if we could make a connection, can the test equipment operate at the frequency? For example, consider a 10 p capacitor, a tiny value that we could get between two little pieces of wire next to each other. At ultra high frequency (uhf) this capacitance has a reactance of only 40 ohms! Similarly, the inductance of conductors with 'no resistance' at audio frequencies becomes greater and greater until, at microwave lengths, 'conductors' become 'insulators', if you see what I mean.

123

The other thing about rf is that it is always trying to radiate, to escape. Where audio is content to be trapped in its wires where it can move about with little restriction, the higher frequencies 'hate confinement', they would rather fly.

Field fault finders will not tackle rf circuitry (if it can be avoided). As student TV-repair-men-to-be, we were warned that the punishment for altering the tune of the IF strip was realignment: we would be forced to re-align it! A time consuming chore requiring specialized equipment and a certain acquired 'knack'. The 'knack' is not something learned once and for all but is a peculiarity of every product and its associated alignment equipment. It will not be mentioned in the manufacturer's tune up procedure probably because it does not look good in print, e.g. 'adjust for maximum (if there is no maximum then adjust for minumum instead)'. Actually this is a reasonable tune up instruction (adjacent coils could be tuned up on the wrong harmonic).

One alignment task that is easy and fun to do is medium wave receiver's IF strip and local oscillator tracking which only requires a 455 Hz source, a suitable trimming tool (for the IF cores) and your ears – find an ancient receiver to practise on (with large chunky quality components) and look up the procedure under 'domestic' mw am rx alignment – at your local library.

Part Two

Fault finding

7 Digital fault finding

Introduction

There are many admirable books written about fault finding microprocessor systems using specialized tools: logic analysers, in-circuit-emulation, fault finding software, etc., but these methods are usually used in development only.

The development engineer must have intimate contact with his system while he debugs both his hardware and software, and these tools provide just that.

In production, the fault finding engineer's problems are different. The design is finished and known to work. It is no longer necessary to know exactly how the software works. The faults will lie in the hardware only.

While it is possible for a fault finding engineer to use development tools for fault finding, it is not usually done. In practice it is far quicker to learn about the hardware only and use the oscilloscope to look into the general functioning of it.

There is a fundamental difficulty in fault finding microprocessor based systems which is best understood from a comparison with analogue electronics.

The television, radio and tape recorder, to name some analogue examples, are primarily amplifiers. A small signal goes in at one end, and a much larger one comes out at the other. In a television, the aerial receives signal strengths of the order of micro volts and produces sound and video of the order of tens or hundreds of volts.

The method for fault finding such an arrangement is to start in the middle of the signal path and look for some amplified portion of the original signal. If that signal is present then, of course, we must look somewhere further along that path towards the output, if that signal is missing or not the expected amplitude or bandwidth then we must look more towards the input and so on. The analogue fault finding method depends on two things:

1 a single signal path;
2 a predictable signal.

In contrast, a processor system

1 has many signal paths, some bi-directional;
2 the system's behaviour is 'unknown'.

This major difference led to a new approach in fault finding, a highly specialized technological one. This separated the fault finder from the machine often with bad results, for example: the engineer has a system that is unreliable. It sometimes works well for days or sometimes it crashes every few minutes. A logic analyser indicates a problem with ROM (read only memory) sometimes a byte fetched from ROM has the wrong value, sometimes the same byte is fetched again with the correct value. When the byte is wrong it is not necessarily the same bit, it might be more than one bit at once. The engineer replaces the plug in ROM and everything seems fine, then a month later the same thing happens again. Eventually another engineer is called in, this one does not know how to operate a logic analyser, or how to write machine code programs, but he has experience, 'feel' and a dual trace scope and he solves the problem in two minutes. What the second engineer has that the first lacks is closer contact with the system and a simpler approach.

128

The 'impossibility' of fault finding a processor system by conventional analogue methods implies the complex route, but in practice, the actual faults that happen in a processor system are generally not digital, but analogue (wrong device fitted, or fitted incorrectly, u/s device, partially u/s device, mechanically intermittent contact, data or address or control line s/c to something or o/c, etc.), truly digital faults hardly ever occur (RAM (random access memory) or ROM fault, processor fault, logic gate delivering a different logic function?). 'Digital faults' occur all the time during development and then the specialized equipment, in the form of circuit emulation, logic analysers, etc., are invaluable in time saved but the development engineer and his equipment would be impracticable for production fault finding.

Brief sketch

The word computer was a poor choice of name because it implies a calculator. A better name is processor, a device that 'processes' data, a simple brain.

The processor, or microprocessor to give it its full name, is totally helpless on its own, it needs contact with the 'outside world' to stimulate it and to allow it to respond by driving some output device(s). The stimulus might come from us as we operate a keyboard, or it might come from another computer via a serial link, etc. The output might be the display on our VDU (visual display unit – after 'video processing') or it might be a serial stream of data to another computer or printer, etc.

There used to be a regular cliché 'computers never make mistakes' but I doubt if any youngsters even know of it. Instead we have a new one, 'to err is human but to mess things up completely needs a computer' which may even be out of date already, as new technology allows easier backing up and more sophisticated programs which entail easier, more intuitive human interfaces and hence less GIGO (Garbage In, Garbage Out).

129

I was going to demonstrate how computers make mistakes, e.g. get your computer to divide 1 by 9 (0.1111111) then multiply the result by 9 (0.9999999) but my calculator gets it correct. Evidently it is rounding up or something. If yours returns the correct answer too then try several divisions of 9 and then the same number of multiplications and it will eventually get it wrong, ever so slightly. Evidently no machine can store all the digits of a recurring number and so it must either round down or round up. I like this 'failing', it gives the computer character and makes it more human. Arithmetic done on one computer may not give quite the same result as the same arithmetic done on another computer of a different make!

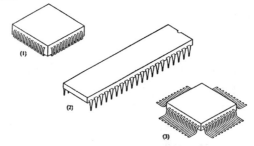

Figure 7.1 (1) J leaded, (2) DIL (dual in line), (3) Gull leaded or flat pack.

Processors come in many different packages (Fig. 7.1). The dil package is the older standard while the other two are surface mounted. Most modern products will be surface mounted and as most of the packages may be physically similar, it will not be immediately apparent which one(s) contains the processor(s).

The processor alone is entirely useless, it must have a clock to drive it and a reset control to put it into a known state. After a reset the processor will start processing, this action is described later on, and it always appears daunting or difficult but believe me, there is nothing to it (from the user's point of view).

The processor's mechanism (architecture) is defined in the same certain way that the moving parts of any machine are defined, e.g.

like the mechanism of an automobile. But like an automobile, it must have guidance. If started up and left to run alone, both the processor and the automobile will crash, they must have instructions to follow.

The processor actively seeks, and expects to get, instructions 'from outside'. These instructions are the program that has been carefully 'written' beforehand and stored in ROM. The motorist also works to a prepared program 'written in advance', e.g. start stationary engine, turn ignition key and release when engine fires.

The analogy breaks down here because although every instruction is written out explicitly for the processor, including all the decisions, it is the processor that actually does the test for the decision. (Evidently the motorist decides if the engine has started.) Once the processor has executed the test, it 'passes' the result back to the program which is responsible for what to make of it. This is where 'bugs' – i.e. errors – in the program may exist. For example, the motorist's program above might work well for years without any problems but one day the engine may not start, then the above routine might continue until the battery goes flat. A superior program would include a 'timeout' – i.e. abandon starting routine after two minutes if unsuccessful – before continuing with a succession of self-tests and their resolutions – i.e. check fuel, if none then fill up, etc.

The processor is able to select an instruction from ROM by manipulating its address bus, asserting its read control, and reading in the instruction via the data bus (this procedure is explained later).

Processors use a group of wires called a 'bus' to connect to their support ics. The processor in Fig. 7.2 has an address bus 20 lines 'wide', labelled A0 to A19, and a data bus 16 lines wide, labelled D0 to D15. The width of the data bus is a measure of the processor's 'power' to compute. Wider data buses allow bigger numbers to be manipulated at one time implying more power. Thus a processor is often 'rated' by its data bus. This one is a 'sixteen bit processor'.

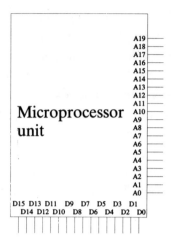

Figure 7.2

There are also 4 bit and 8 bit processors which are less powerful as well as more powerful processors with data buses as wide as 32 bits or more. The size of the processor has no bearing on the following fault finding methods and as large buses only add confusion, the rest of this book only cites 8 bit examples. The methods apply to any processor type irrespective of instruction set, or data bus width.

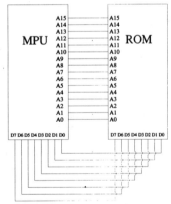

Figure 7.3 An 8 bit processor connected to its ROM.

The processor connects to its ROM with its address and data lines as illustrated in Fig. 7.3. Now the processor is able to read and then follow the program contained in the ROM but we still need a way for the processor to interact with the outside world so that it can do something useful. This device is termed an input/output device (IO device) but before we get in too deep too early let's take a look at memory (ROM).

8 Memory

There are various types of memory used by processor systems to store programs and data, e.g. floppy discs, hard discs, tape streamers, etc., but for our purposes, we need to concentrate on that type of memory which the processor has direct access to, i.e. ROM and RAM (direct access means access via the address and data buses).

RAM (Random Access Memory) and ROM (Read Only Memory) differ in only one way, which is that RAM can have its contents altered at any time whilst ROM cannot, the ROM's contents are fixed beforehand and are retained permanently even without power. The RAM will only hold its contents while power is applied to it. Both RAM and ROM are 'random access memories', this phrase was meant to distinguish their nature from magnetic tape memories (i.e. to access data from tape is slow and variable, if the data you want is at the other end of the tape it could take ages to wind it through). With a ROM or RAM ic it always takes the same time to access any data whether it is at the start, end or middle of the ic. Therefore any data selected at random takes the same time to access (access times of 100 ns are commonplace).

ROM with two locations

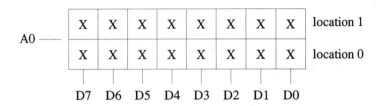

Figure 8.1

Each location in Fig. 8.1 contains eight bits. A bit (labelled X) is either a one or zero. The bits may be connected to the data bus. These connections are made eight bits at a time (eight bits are called a byte). If we call the bottom location zero and the top location 1 then either the bottom eight bits can be connected to the data bus or the top eight bits can. This process is called addressing and is performed by the address bus.

This address bus has only one wire labelled A0. Can you see why? It is because only one wire is needed to indicate a zero (low) or a one (high), the zero would select location zero and the one would select location one. A0 can be either high or low: high is +5 V; low is 0 V.

If we had two wires (A0 and A1) then we would be able to make four different selections like this:

A1=low A0=low

A1=low A0=high

A1=high A0=low

A1=high A0=high

We could call these selections zero, one, two and three like this:

A1	A0	Selection
0	0	0
0	1	1
1	0	2
1	1	3

RAM

RAM is identical to ROM with one exception: the read/write control. When we wish to change a location in the RAM we first select the location (set A0 to +5 V or clear it to 0 V) then we set the read/write control high (to read the selected location on to the data bus) or clear the read/write to low (to write the eight bits of data on the data bus into the selected location). So the diagram of a two location RAM would look like Fig. 8.2.

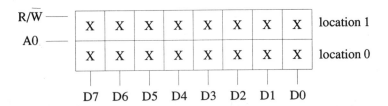

Figure 8.2 RAM with two locations.

Do you understand everything so far? Be sure that you do, don't be tempted by this slow pace to rush ahead prematurely. Everything will get more complicated quite soon enough! Now try the following exercises.

Exercises

1 Draw a block diagram of a 4 location RAM eight bits wide.
2 If location 0 was selected what voltage would be on A0 and A1?
3 If A0 and A1 were both low what location would be selected?
4 If the read/write line and +5 V supply were held permanently high then what would be the difference between this 4 location RAM and a 4 location ROM?

Answers:

1 See Fig. 8.3

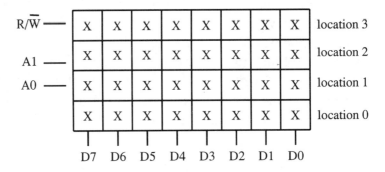

Figure 8.3

2 A0 and A1 would both be 0 V.
3 Location 0 would be selected.
4 No difference, the processor would simply see four locations with permanent, unchangeable contents.

9 Binary and hexadecimal

The binary number system might seem hard to understand. This is because we have accepted our decimal system as the natural way of counting. If we take a fresh look at decimal numbers, then the binary number system may not seem so very different after all. Both systems are simply ways of grouping things. The decimal way is to group things in tens, the binary way is to group in twos. We humans have decided that groups of tens are easier to imagine and simpler to write down. Digital computer systems are simpler to engineer using the binary system.

If we spread some things randomly on this page:

```
** ** *      *   * **
  *** * ** **  * *
 *** ** * *  * *  *  *  * **
```

it is too hard for us to say how many things there are, but if we group these things in tens so:

```
********* ********* ********* ***
```

then it is easy to tell at a glance how many there are; three groups of ten and three left over is 33. Here we have decided that the biggest number we wish to see is nine, that is, if we have more than nine things then we must start grouping them in tens, making for easier counting. If we continue counting and find that we have more than nine groups of ten, then we group these groups into tens, that is groups of ten tens so:

```
*********   *********   *********               *********  *********  **
*********   *********   *********
*********   *********   *********
*********   *********   *********
*********   *********   *********
*********   *********   *********
*********   *********   *********
*********   *********   *********
*********   *********   *********
*********   *********   *********
```

Almost at a glance we can see that the number is 322, i.e. we now have three sets of ten tens, two sets of tens and two. This repeated grouping of tens means that we only need to invent the ten unique number names, zero through to nine.

A binary grouping is difficult to visualize, e.g. how big is the binary number 101001100? And is this bigger than 100110101? And if so by how much? To make binary easier to visualize and to convert to decimal, we use another number system called hexadecimal.

Hexadecimal

This system is used when we talk computer numbers. It is a halfway house between the easy-for-humans decimal system and the hard-for-humans binary system. We will be able to convert hex to binary in our heads, and vice versa, after a little practice.

Hexadecimal counts from 0 to 9 like decimal:

decimal	hexadecimal
0	0
1	1
2	2
3	3
4	4
5	5
6	6
7	7
8	8
9	9

Until we get to ten, then the systems differ:

decimal	hexadecimal
10	A
11	B
12	C
13	D
14	E
15	F

That is, in hex, the units are from 0 to 15, called 0 to F. If you have understood the decimal system above you might guess what happens next when we count the next number 16:

decimal	hexadecimal
16	10

That is, in hex the numbers are grouped in sixteens. To continue:

decimal	hexadecimal
17	11
18	12
19	13
20	14
21	15
22	16
23	17
24	18
25	19
26	1A
27	1B
28	1C
29	1D
30	1E
31	1F
32	20

If we were told that a hex number was 3F we could work out what it is, like this: the three means three groups of sixteen with F left over (fifteen left over) so that must be 48 plus fifteen, equalling 63:

$$63 = 3F$$

If we had a hex number of 100 (called 'one zero zero', because 'one hundred' is a decimal number) we would work it out as one group of sixteen sixteens with no units, one times sixteen squared:

$$1 \times 16^2 = 256 \text{ that is:}$$

256 in decimal = 100 in hex

Can you see what we are doing? Each column of a decimal number has a 'weight' of a power of ten and each column of a hexadecimal number is a power of sixteen:

decimal				hexadecimal		
	256		=		100	
2	5	6	=	1	0	0
2×10^2 +	5×10^1 +	6×10^0	=	1×16^2 +	0×16^1 +	0×16^0

Binary

The binary system has only two numbers: 0 and 1. Decimal counting and binary from 0 to 1 is identical:

decimal	binary
0	0
1	1

Then for the count of two:

decimal	binary
2	10

That is, units are grouped in twos so for three units the binary arrangement has one group of two and one over:

decimal	binary
3	11

and so on:

decimal	binary
4	100
5	101
6	110
7	111
8	1000
9	1001
10	1010
11	1011
12	1100
13	1101
14	1110
15	1111

Have a look at the binary progression and note how:

a All even numbers end in zero.
b All odd numbers end in a one.
c Look at the way 7 progresses to 8, can you see how the binary addition works? If you can, then you will be able to predict the next number:

decimal	binary
16	10000

Did you get it? If not, look at this decimal addition:

$$9999$$
$$1+$$

$$10000$$

It works in the same way as this binary addition:

$$1111$$
$$1+$$

$$10000$$

Hex and binary

Conversion from hex to binary and vice versa is simple once the following binary numbers are memorized.

hexadecimal	binary
0	0
1	1
2	10
3	11
4	100
5	101
6	110
7	111
8	1000
9	1001
A	1010
B	1011
C	1100
D	1101
E	1110
F	1111

You may find it easier to remember a few 'key' binary patterns and from these work out the others, i.e.:

binary
100
1000
1100

If you know that eight is 1000 then seven must be one less, i.e. 111 and nine is one more, i.e. 1001. If you know that twelve is 1100 (C) then (D) must be one more 1101 (thirteen), etc.

Notice that all the binary numbers from 0000 to 1111 fit into four columns, while the equivalent hex numbers fit into one column, this rule makes conversion from one to the other simple:

hexadecimal value FFFF = binary value 1111111111111111

If we split the binary into groups of four bits:

hex	=	F	F	F	F
binary	=	1111	1111	1111	1111

What is 3A78 in binary? Easy, the 3 is 0011, the 'A' is 1010, the 7 is 0111, the 8 is 1000:

hex	=	3	A	7	8
binary	=	0011	1010	0111	1000

What is 1011000011001011 in hex? 1011 is B, 0000 is 0, 1100 is C, 1011 is B so:

binary	=	1011	000	1100	1011
hex	=	B	0	C	B

Exercise

1 One trace of an oscilloscope is triggered on A15 while the other trace is A14. What address range is the processor accessing (Fig. 9.1)?

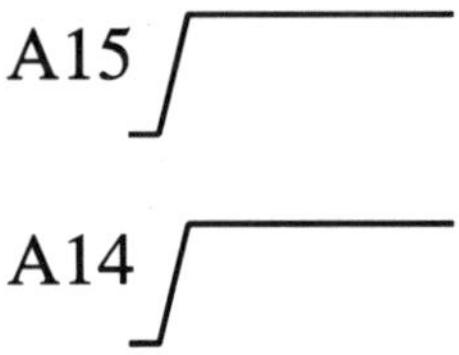

A15

A14

Figure 9.1

2 The second trace is moved to scope each data bit and the read/write line in turn, what is happening during this address range (Fig. 9.2)?

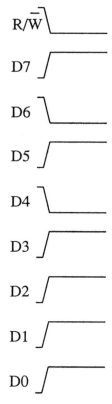

R/W̄

D7

D6

D5

D4

D3

D2

D1

Figure 9.2

D0

Answers:

1 Because A15 is high the address must be somewhere between 8000 and FFFF, A14 is also high so the range is narrowed to between C000 and FFFF.
2 The read/write bar line is low so it must be a write cycle, the data bits from D7 to D0 are 10101111 = AF therefore the processor is writing AF to something in the C0000 to FFFF range.

10 Addressing

Chapter Eight (Memory) introduced a two location memory selected by one address line. If two address lines are used then four locations can be selected. Every time a new address line is introduced the address range, or field, is doubled:

A0	1	address	line can select	2	locations
A1	2	"	lines can select	4	"
A2	3	"		8	"
A3	4	"		16	"
A4	5	"		32	"
A5	6	"		64	"
A6	7	"		128	"
A7	8	"		256	"
A8	9	"		512	"
A9	10	"		1024	"
A10	11	"		2048	"
A11	12	"		4096	"
A12	13	"		8192	"
A13	14	"		16384	"
A14	15	"		32768	"
A15	16	"		65536	"
A16	17	"		128K	"
A17	18	"		256K	"
A18	19	"		512K	"
A19	20	"		1M	"
A20	21	"		2M	"

A value of 1024 equals 1 k in computer jargon so 64 k means 65 536. 1 Mbyte means one megabyte or 1 048 576 bytes.

To calculate the address field, raise two to the power of the total address lines, e.g. for sixteen address lines there are 2^{16} or 64 k locations.

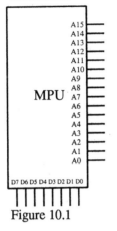

How many locations can the processor in Fig. 10.1 address directly? It must be 64 k because A0 to A15 = sixteen lines.

Figure 10.1

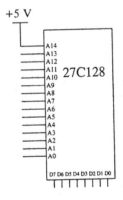

Figure 10.2

What is the maximum accessible size of the program in the eprom in Fig. 10.2? It must be 8 k. The eprom size is evidently 16 k but because the top address line is fixed high, only the top half is accessible (if this is not clear it is explained again below). The type number 27C128 also indicates the size. The 128 means 128 k bits, i.e. 128 k/8 = 16 k bytes and the 'C' indicates CMOS technology.

Examine the binary count below. Note how the units change for every count, how the 2s change on every second count, how the 4s column changes for every count of 4 and so on.

150

0	0
1	1
10	1x
11	1x
100	1xx
101	1xx
110	1xx
111	1xx
1000	1xxx
1001	1xxx
1010	1xxx
1011	1xxx
1100	1xxx
1101	1xxx
1110	1xxx
1111	1xxx
10000	1xxxx
10001	1xxxx
10010	1xxxx
10011	1xxxx
10100	1xxxx
10101	1xxxx
10110	1xxxx
10111	1xxxx
11000	1xxxx
11001	1xxxx
11010	1xxxx
11011	1xxxx
11100	1xxxx
11101	1xxxx
11110	1xxxx
11111	1xxxx

Half of the addresses (zero to 15) have A4 low while the other half (16 to 31) have A4 high. If we wished to cut the 32 location area in half we would use the top address line (A4). In the processor example in Fig. 10.1, if we were to monitor the state of A15 (the top line) we could tell if the processor were accessing either the top half (A15 high) or the bottom half (A15 low) of its address field.

Addressing

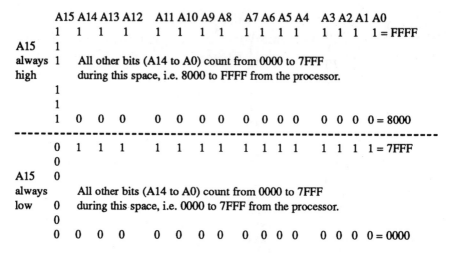

Figure 10.3 Top address line divides field into halves.

In the example in Fig. 10.3, if we were connected to A0 only, we would have no idea where we were in the address field, we would only know odd from even. If we were connected from A14 to A0 we would know our exact position in either the top half or the bottom half of the field, but not which half.

Memory mapping

If a processor has sixteen address lines then it can scan an address field of 65 536 (64 k) different locations. We could draw a map of those locations as illustrated in Fig. 10.4.

Figure 10.4 A 64 k memory map.

152

The bottom location is labelled 0000 and the top FFFF. If we connected a 64 k byte memory then all the locations in that memory would fit into this map exactly. The circuit diagram would look like Fig. 10.5.

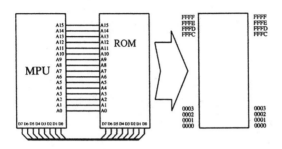

Figure 10.5

If we connected a 2 byte memory to the same processor (as in Fig. 10.6) then, if the processor read address = 0000, it would see location 0 of the memory; if it read address 0001, it would select location 1. What would it see if it read location 0002? It would see location 0 again, likewise when it read location 0003 it would see location 1. In this way the processor would see the same 2 byte

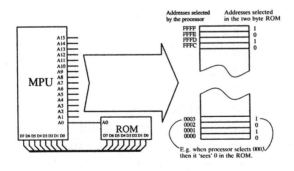

Figure 10.6

153

memory throughout its 64 k byte address field 32 k times over. If that memory were RAM memory, then if the processor were to write a value into location 0000, it could read that new value at location 0000 or location 0002 or location FFFe or any even location.

If we connected a 32 k memory to the processor's address bus, then all locations from 0000 to 7FFF (the bottom 32 k of the field) would be read from that memory (Fig. 10.7).

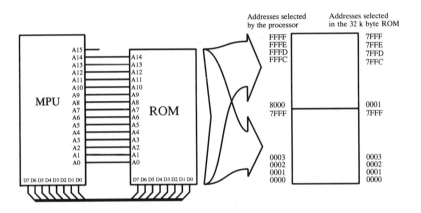

Figure 10.7

If the processor tried to access location 8000, then all address lines to the memory A0 to A14 would be low, that is the memory would give the contents of its 0000 location, if the processor tried 8001 the memory would give 0001 and so the memory would appear in the processor's address field again, that is from 8000 to FFFF (the top 32 k) would be read from the memory's 0000 to 7FFF locations. The complete map, as the processor sees it, would look like Fig. 10.8.

A processor with sixteen address lines should be able to access two 32 k byte devices in its 64 k byte field but we cannot simply connect them both to the processor's buses like this: why not? That is, why will the configuration in Fig. 10.9 not work?

154

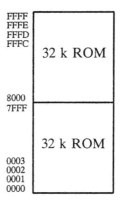

Figure 10.8

If the processor tried to read location 0 in one of the ROMs, then that ROM will output its data on to the data bus and each data line will be driven low or high by that ROM: meanwhile the other ROM will also try to output its data on to the same data bus and the two will clash (unless the two bytes are exactly the same). To prevent clashes of this sort (bus contention), each ROM must be disconnected from the data bus when not in use. This is done with the chip select (chip enable) and the output enable controls.

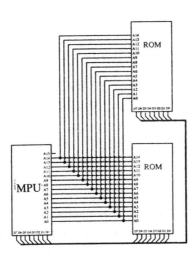

Figure 10.9

155

All devices that share the data bus must have some means built into them that allow their data outputs to be turned off. When a device is chip selected (or enabled) then its outputs will drive the bus, i.e. they must go low impedance. When a device is deselected (or disabled) then the outputs are disconnected from the bus, i.e. the outputs go high impedance.

This high impedance condition is the so-called tri-state, i.e. an output can be low or high or floating. Of course, as we know already, in impedance terms there are only two states: low impedance when the chip is selected and the output is either high or low, and high impedance when the output is floating.

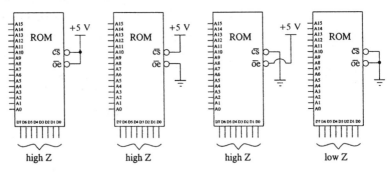

Figure 10.10 The bar and the circle symbols mean *active low*.

For example, the devices in Fig. 10.10 only drive their data buses when both the chip select and output enables are driven low. Conversely a control line may be active high (which is implied by the lack of the bar and the little circle), but beware, these inversion symbols are often omitted accidentally. The bar always comes after the name, e.g.:

- 'read bar write' means that the control is active low for 'read';
- 'read write bar' means that the control is active low for 'write';
- 'bar read write' has no meaning.

156

The general term for active is 'asserted'. If a control is asserted then it must be in the active state (whatever that happens to be, whether high or low) and inactive if de-asserted.

If this is new to you then take your time and be sure to learn the conventions otherwise it is easy to get muddled as things speed up!

Address decoding

Both 32 k memories in Fig. 10.11 must share A0 to A14 leaving A15 redundant, what does A15 mean? If we again look at all the addresses from 0000 to 7FFF we see that for all of them A15 is low, while all the addresses from 8000 to FFFF (the upper 32 k of the address field) have A15 high. That is A15's binary 'weight' is 32 k ($2^{15} = 32$ k), just as A0's weight is 1 ($2^0 = 1$), so just as A0 signifies groups of 1, so A15 signifies groups of 32 k. If we take A15 and use it as a switch to select either memory on or off then we can arbitrarily create a memory map as shown in Fig. 10.11.

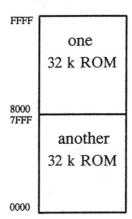

Figure 10.11

If a circuit arrangement like the one illustrated in Fig. 10.12 is used which ROM is in the top half of the address field? (*Note:* CS is asserted positive.)

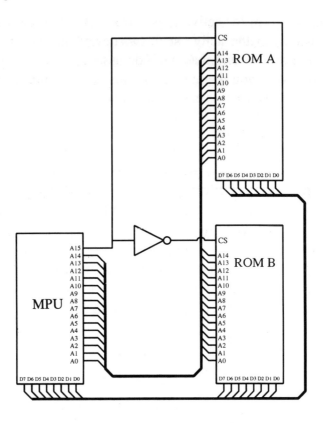

Figure 10.12

Answer: When A15 goes high, ROM B will be deselected (the inverter will output a low into chip select) and ROM A will be selected (high into chip select), i.e. only ROM A will appear in the higher range of 8000 to FFFF.

When A15 goes low, then ROM B will be selected and ROM A will be deselected, i.e. only ROM B will appear in the lower range of 0000 to 7FFF.

158

We could fit four 16 k memories into the processor's address field as demonstrated by Fig. 10.13, using A15 and A14 to drive a two into four line decoder whose truth table works like Fig. 10.14. Can you work out where each ROM sits in the memory map?

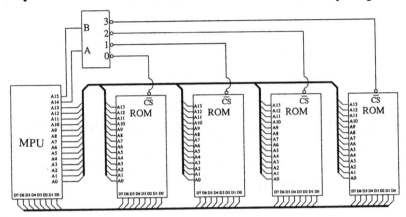

Figure 10.13

A	B	0	1	2	3
0	0	**0**	1	1	1
1	0	1	**0**	1	1
0	1	1	1	**0**	1
1	1	1	1	1	**0**

Figure 10.14 Two into four line decoder: when both A and B are low then output 0 is selected and because the device above is active low, then 0 must go low. Similarly, when A is high and B is low then output 1 is selected.

Answer: The ROM selected by output '3' will turn on for any address with *both* A15 *and* A14 *high*, i.e. any address between C000 and FFFF, i.e. the top quarter of the map. The ROM selected by output '2' will turn on for any address with A15 *high and* A14 *low*, i.e. any address between 8000 and BFFF, i.e. the next quarter of the map (Fig. 10.15).

159

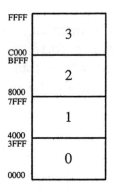

Figure 10.15

Similarly 'ROM 1' sits at addresses 4000 to 7FFF and 'ROM 0' sits at the bottom, 0000 to 3FFF.

Suppose we view A15 and A14 and see a configuration as in Fig. 10.16. What is the processor doing?

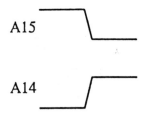

Figure 10.16

Answer: It is first selecting ROM 2 then selecting ROM 1.

Input output (IO)

We could modify the field to accommodate an output device (e.g. an LED – light emitting diode) as in Fig. 10.17. It would appear as in Fig. 10.18 in the processor's map. To turn on the LED the processor would access any address in the range between C000 and FFFF. Can you see why this would be unsatisfactory?

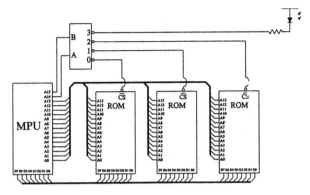

Figure 10.17

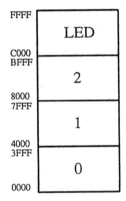

Figure 10.18

Answer: The program that turns on the LED must live in a ROM. The processor would be forced to switch between that ROM and the LED. This would occur at system speed (μs) and the LED would shine for only one mpu cycle at a time.

Should we wish to control an output bit, we would need a little more circuitry to allow the processor to latch the bit high or low (ignoring the waste of 16 k of address field to control a single bit!)

161

11 Discrete logic

Twenty years ago a 'control board' would probably have been built entirely of 'discrete' logic (individual ics such as quad two input NANDs, dual d-type flip flops, shift registers, etc.)

These days discrete logic is always minimized as designers try to reduce the size, complexity and cost of units by implementing most of the control task in firmware. Nowadays the few discrete logic packages dotted around the microcontroller are the faint remnants of a glorious past!

At first AND and OR gates, NOT and exclusive-OR may seem a little abstract, especially when grouped together in a fault finding exercise. It is not so difficult. The trick is to decide on-the-fly which input(s) are 'overriding' and concentrate on them only in order to minimize scope readings.

In reading the following notes it will become apparent that we are not testing all the logic states. That is not the objective, we do not wish to test the circuitry, for that is a completely different process, instead we shall be proving a single or a few conditions to serve our immediate purpose.

'Dominating' effects

To check a combination of logic gates expediently, learn to recognize dominating states, e.g. for the NAND gate this is any input low (Fig. 11.1).

Discrete logic

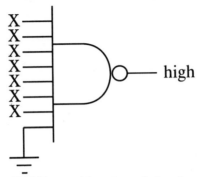

Figure 11.1 8 input NAND gate, if *any* input is *low* then the output is forced *high*.

Any single input high will not determine the output, but if any single input is low then the output will be forced high. When fault finding, check each input for the dominant low condition, if one is low then check that the output is high, ignoring any other inputs; they are irrelevant. Evidently, if all inputs are high then the output should be low.

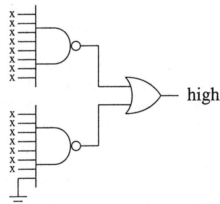

Figure 11.2 NAND OR combination, a low on *any* input will force the output *high*.

The dominant condition for the OR gate input is high. A high on either of its inputs will be forced by a low on any NAND input. Therefore a single low on any NAND input will force a high output.

164

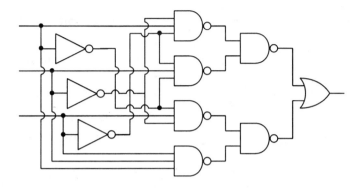

Figure 11.3

NAND OR combination:

a A high input to the OR will dominate.
b Any low input to the 2 input NANDS will dominate.
c To achieve (b) all inputs of any three input NAND must be high.

If any of those inputs (in Fig. 11.3) are low then they dominate that NAND forcing us to move onto the next 3 input NAND and so on. Only if one of those NANDS has all three inputs high do we bother to check the OR op (for high). On the contrary, if all of those NANDS have at least one input low then we test the OR output for a low.

To relate all this to the fault finding that is to come: suppose we compare contention-on-the-data-bus (this concept is explained later) with some discrete logic ics that have something to do with controlling access to that bus. An error may show up in the operation of the discrete logic which is easier to spot by looking for and testing dominant conditions.

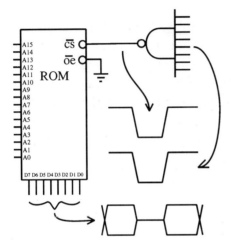

All data bits (D7 to D0) contend during a particular cycle. During this contention the ROM is chip selected by the NAND output which should be forced high by the 'dominant low' on one of its inputs. In this case the NAND is faulty. Do not worry about 'contention', this will be covered in detail later.

Figure 11.4

Exercise

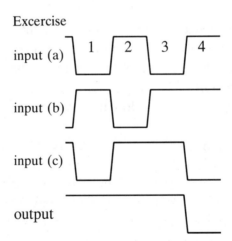

Figure 11.5 Three input NAND gate: input and output signals.

What is wrong in Fig. 11.5?

Answer: The input (c) is o/c. During cycle number 4 input (c) should force the output high but the output goes low, therefore the device thinks that input (c) is high. If the above waveform was measured at the ic's pin, then the package is u/s. If the above waveform was measured at an output supposedly driving that gate then the continuity to the pin could be broken; if that device was a transistor-transistor-logic (TTL) package (74LS10) a disconnected input would float at 2 V, a high level.

Technologies

TTL

Early integrated circuits were constructed from a technology called transistor-transistor-logic (TTL), the '74 series'. A device with the number 7400 is the first of this series which always starts with the prefix '74'. When they first appeared, the speed at which these devices operated was considered very high. Engineers had difficulty following a pulse through a logic circuit. To facilitate fault finding a common practice was to use the logic probe, a simple device that would 'stretch' a short pulse of a micro second or less to a second or more, long enough to drive an LED visibly.

The 74S00 is a schottky version of the 7400. The speed of a schottky device is roughly ten times faster than a 74xx. This means that a faulty 74Sxx device should *never* be replaced with a 74xx even if it seems to work, the long-term problems that might occur could be difficult for another engineer to find. A drawback of schottky is the higher power demanded, but from it sprang an improved version of the 74xx; low power schottky. Designated 74LSxx, low power schottky has a similar speed to the 74xx but consumes less power. As a result it became the preferred general choice. Other than the schmitt triggered, all these technologies (of the TTL family) have the same characteristics:

a An o/c input floats at 2 V.

b The output drive is asymmetrical, TTL pulls down harder than it pulls up. This is because the output stage has a single pull down transistor and only a collector resistor for pull up.

If a 2V dc level appears anywhere in a TTL system, something is wrong. If that 2 V level is floating (can a finger rested on the scope probe 'pull' it, or an adjacent TTL drive it without contention?), then the drive stage to it must be u/s or the electrical connection broken. To make this clearer consider the readings in Fig. 11.6.

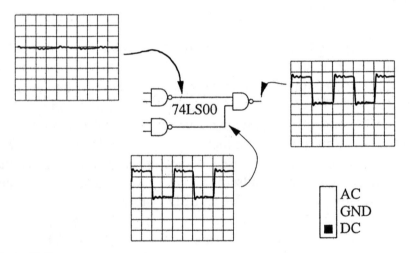

Figure 11.6

The *Y* amp is directly coupled (which means that 0 V is positioned on the centre line). As logic voltages swing between 0 V (low) and +5 V (high) the *Y* amp must be set to 2 V per division. This makes the voltage of the left hand trace about +2 V dc which is illegal! Note that this level is rather noisy. The noise is pickup from adjacent tracking which is a clue: what does it mean?

Answer: It means that the point we are looking at is high impedance. If it were low impedance, pickup would be minimized (shortened to ground through the low impedance).

If this point is high impedance then what has happened to the NAND output that is supposed to be driving this line? It is either missing because the tracking to it is broken or it is destroyed.

Now at every circuit node there is always some pickup so this diagnosis so far is rather suspect. We must be more certain that this node is high Z. This is where the 'finger test' comes in (Fig. 11.7).

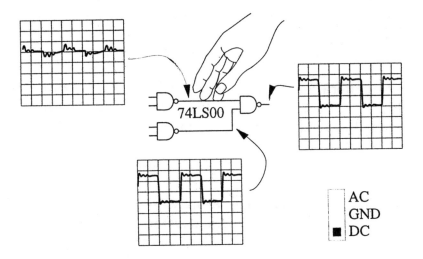

Figure 11.7

Touching the suspect node will either have no effect on the trace or, as shown above, the dc level will be easily displaced and extra noise and possibly line hum will be injected which proves this node is high impedance (floating), i.e. the preceding NAND output is u/s or not connected.

If the finger test had had no effect what would that imply?

169

Answer: That the node is low impedance. There are four possibilities, either:

1 The preceding NAND output is destroyed leaving it permanently outputting 2 V dc with some force, i.e. low impedance.
2 The following NAND input is destroyed with an identical symptom, i.e. low impedance 2 V dc 'output'.
3 The following NAND input is destroyed so as to load the preceding NAND's high output to ground (or vice versa, i.e. loading the preceding NAND's low output upwards).
4 Some other independent node is s/c to this one, driving it hard to 2 V dc: 'swamping' any traces of signal.

Another effortless method to prove whether the above case is low or high Z is to momentarily s/c the 00's input pins with the tip of the probe and look for 'contention': if the probe is currently on pin 1 (say) then while maintaining contact with pin 1, slide the probe into contact with pin 2 (the other input of that gate) and slide it away again.

The trace will either change entirely from the staid 2 V dc level to the dynamic 'correct' 0 V to 5 V waveform and back again (faulty node is therefore floating) or it will remain solidly at 2 V dc without any 'pulling' effect from the brief coupling to the active signal on pin 2 (faulty node is therefore low Z).

Incidentally, if the latter happened then we should also repeat the experiment from the other viewpoint to be absolutely sure that both nodes were actually s/c together, i.e. whilst maintaining contact with pin 2 slide the probe into pin 1 and back again. Evidently this time we would expect to see first the dynamic level, which would then be 'overridden' by the 2 V dc level, during the brief short, and then the dynamic level should reappear as pin 1 disconnects.

Advice

If you do not already know what an '00' is, then look it up and memorize it, also try to get familiar with other common types, i.e. 04, 14, 74, 161/2/3/4, 240/4, 245, 259, 373/4, etc. It is not so difficult. Most logic packages have the supply pins 'at opposite corners', so for the 7400, 74LS00, 74ALS00, 74HCT00, 74S00, and 74HC00 devices, to name some of the '00' variants, supply will be pin 14 and earth pin 7. The only thing you have to learn is 'quad NAND – all pointing downward', i.e. pins 1 and 2 must be input and three the output. Pins 13 and 12 must be inputs with 11 being the output, etc. (Evidently they must be two input NANDs otherwise four of them would not fit in a 14 pin package). Knowing these things is some protection from wrongly labelled diagrams and is the sort of knowledge which can make all the difference at a job interview.

CMOS

The early '4000 series' CMOS integrated circuits, and the '74 series' TTL ics, are quite different, i.e. apart from the technologies, similar part numbers have different logic functions. This contrasts with the modern 74HC series CMOS whose logic functions and pinouts are identical to the 74 TTL series.

Today most TTL devices can be directly replaced with CMOS equivalents. A pcb that would have been built with 74LSxx will now almost certainly be exclusively built with 74HCxx.

CMOS, both the older 4000 series and the 74HC, have the following characteristics:

a It needs such little energy at low speed that it can 'live on air'. If the supply (V_{dd}) to a package is missing, it can still 'work'. As the engineer probes closer to the problem the scope will increasingly influence the surrounding circuitry, exacerbating the problem until scoping the ic itself 'kills' its operation.

171

b Its output is symmetrical, connection between CMOS outputs tends to settle at half the supply (there is more on contention below).

c It has very high input impedance. We cannot be sure when scoping an input if it is low because it is driven low or whether it is floating and pulled low by the scope.

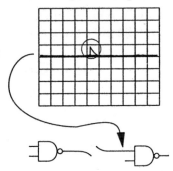

Figure 11.8 At the instant that the scope probes the floating input, there may be an extremely faint 'vestige' of exponential decay as the scope pulls the input level to ground.

An alert engineer might notice the scope's trace transit from high to low exponentially at the instant of connection. (This may not be repeatable unless the unit is powered off and on again.) If so, then confirm with the 'finger test' or the momentary-short-to-an-active-output test described above.

In examples like this, the idea is not to test every node specifically for this or that condition or to switch every piece of equipment off and on! The idea is simply to be aware and ready to take advantage of any 'free gifts' if they crop up during a search for something else! Probably half of all difficult faults are found by 'chance observation', i.e. luck can be said to be proportional to experience.

A fourth characteristic of CMOS integrated circuits is that low frequency, low energy circuits can suddenly increase their current demand many thousands of times if either a CMOS i/p (input) is disconnected or a fault holds a CMOS input within the thresholds. This can turn on every transistor in the package. Every complementary pair will then s/c the supply and ground rails together. This fault can be confusing. The temptation is to 'change the one(s) that is (are) getting hot' when the cause is in fact in the continuity between the packages – this type of problem can occur from pcb damage inflicted by bad rework. The extra demand on the supply might load it down to near ground, making it impossible to trace the fault dynamically. Try turning it off and on – there may be a window of a few seconds before the offending gate's input(s) float into the threshold level – or try breathing on the pcb (or use a freezer) – condensing breath might load a floating input high or low. If none of these work maybe an input is not floating; perhaps an inverter has its input and output s/c together with a resulting high frequency oscillation keeping that package and all others driven from it in a mainly conducting state.

The 74 series and the 4000 series have nothing in common. The 74HC and 74HCT were developed to overcome this. The 74HCT is a CMOS equivalent of a 74 device; in theory both technologies can be freely mixed but as a fault finder always try to replace like with like, even use the same manufacturer if available. The HCT suffix means High Speed CMOS with TTL configured input thresholds. The HC suffix means High Speed CMOS with CMOS configured inputs. HC can run on supplies as low as 2 V up to 6 V, the HCT of course is restricted like TTL to 5 V only.

Liquid damage

When servicing returns from the field, watch out for any signs of liquid spillage. They may have no immediate effect when dry but may corrode fine pcb tracking later on when 'reactivated' by condensation.

Mouse's fault

When a particular system was switched on, it would partially initialize, 'hanging up' in the latter stages when the keyboard was accessed (there were start up options which could be preset by switches on the keyboard).

Changing the IO device (system bus to keyboard interface) had no effect. A cursory glance around the IO's control lines found the reset line sitting at 1 V, evidently the IO device was held in a reset condition. There were two resets, one for the entire system and one for this IO device. The IO reset was very simple (Fig. 11.9).

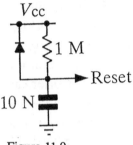

Figure 11.9

Evidently the reset line was loaded down by something. (If the 1 M was o/c then the load of the scope should easily ground this node – otherwise 1 M would not pull up hard enough in the first place.) So the end of the 10 N was lifted and the fault remained, i.e. it was not leaky. As nothing else was connected to reset, apart from the IO device (and this was substituted above), there must be something else connected to this node, but what? A fault like this is possible in 'production' but the machine was a field failure, i.e. free of production style faults, unless the 'something-connected-to-it' had been there all along but had only got 'worse recently' – weird thoughts like this happen to the best of us! The answer turned out to be even stranger.

As I was pondering this particular problem, I noticed out of the corner of my eye that the scope trace was displaced 2 V negative. I had rested my scope hand with the probe tip lying on a copper-free pcb surface, i.e. the board surface was conductive. Nearby was a –5 V feed. The reset tracking also ran through this area.

I cleaned the pcb surface and the pull down effect vanished. While doing this I discovered some little brown pellets about a millimetre long; the customer, who was watching, recognized these as mouse droppings. His machine had an open hole in the back, for a DIN socket, and this was apparently not too small for a mouse. The conductive surface was presumably mouse urine.

174

12 Microprocessor action

Understanding what a microprocessor actually does (not how it actually works, which is irrelevant) can only be had by writing programs for them in 'machine code' or, more accurately, in 'assembler'. These programs are written in the same form that the processor actually 'sees' as it works, called assembly language programs because they must be 'assembled' first to produce the final machine code. The reason for this is that the final code is a string of bytes that, if written in hexadecimal, would look like this:

50
3F
4C
A8
B4
A9

Though it would be possible to memorize what these hex codes mean, it is better to give them apt labels, e.g.:

ldab

which means 'load accumulator b'. A program written using such labels is called the assembly language program, and the act of assembling it would produce a string of binary bytes derived from those labels. These labels, or 'mnemonics', are the instructions that tell the processor what to do next. A complete list of these instructions is the 'instruction set'.

Not all programs are written in assembly language. There are many so-called higher level languages like Basic, C, Pascal, Fortran, etc. The labels they use have higher meanings, e.g. the Basic statement:

PRINT

means 'print something on the VDU'. To do this in assembly language with the same features would require many assembly instructions.

Of course all higher level languages cannot be understood by a processor unless they are converted into machine code. For instance PRINT will be a specific set of machine code instructions which are always used whenever PRINT is encountered, a so-called machine code subroutine.

A high level language is only a formal way of specifying machine code subroutines. An advantage in doing this is that PRINT will mean the same thing for all computers, even though they might use different processors with different instruction sets. The other advantage, of course, is that all the low level routines have been already written so that the programmer can concentrate on the actual job in hand without having to understand anything about the processor or the hardware that it must access. The subroutines used in a high level language are specifically written to suit particular tasks: COBOL is for business use, C is for lower level system functions, etc.

Fortunately for us it is not necessary to understand machine code programming to fault find a processor system but it is necessary to grasp the nature of processing, hence this chapter 'Microprocessor action'.

176

Reset

When power is first applied, the many thousands of elements inside the processor will flip into random states and in this unknown condition the processor would react in an unpredictable way. To initialize it, the reset pin must be held active (for a given minimum number of clock cycles) and released, then a prearranged internal sequence will put the processor in a known state and will 'let it loose'. (The 'starting' address, the 'reset vector', is fixed in the architecture and is always the same for a particular processor type. When the program is written, the actual starting address (which is arbitrary) of that program is placed in this vector. When reset, the processor takes the contents of this vector as the actual start address, hence the name vector, i.e. pointer, an example of what software engineers call 'indirection'.)

Processing action proceeds in a repetitive manner. First a byte or bytes are accessed and read from memory into the instruction decoder. This is called 'instruction fetch'. The instruction is decoded and the processor performs an operation accordingly. For instance, NOP (no operation) means do nothing, in which case the processor will proceed by stepping the address bus to the next byte(s) up and will begin again with an instruction fetch. This 'stepping the address bus to the next location up' is a typical processor action; the address bus is said to be incremented. Some instructions have data attached to them, which is required by the instruction. The processor cannot know in advance that an instruction and some data are on the way, but after decoding, the instruction itself will force the processor to dutifully increment the address bus to gather that data. It will then increment the address bus and begin with an instruction fetch again. An example is given in Fig. 12.1.

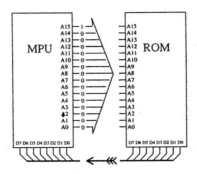

Figure 12.1 Step 1 Instruction Fetch: after incrementing the address bus the processor reads a byte which it assumes is an instruction.

Step 1

Instruction Fetch: the processor has just completed a previous instruction and because of this, it has set itself up for a new instruction fetch. The old address on the address bus was 7FFF, so the first thing the processor does is to increment the address bus to access the next location up. The new address = 7FFF+1 = 8000. Once the new address is set up then the memory is read into the processor. This value, whatever it may be, will be assumed by the processor to be an instruction, because the processor is in its instruction fetch phase (Fig. 12.1)

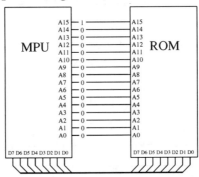

Figure 12.2 Step 2 Instruction decode: the processor decodes the instruction.

Step 2

Instruction decode: the processor decodes the last instruction internally (Fig. 12.2).

178

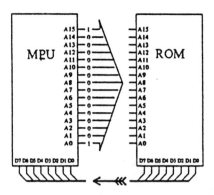

Figure 12.3 Step 3 This particular instruction, after decoding, 'tells' the processor that it wants such and such doing with the following two bytes of data. So the processor must get the first of those following bytes by incrementing the address bus.

Step 3

The instruction has been decoded. The instruction (like most of them) requires extra data (Fig. 12.3). So the processor increments the address bus, 8000 + 1 = 8001 and reads in first one byte and then another, and so on.

(If it had been a 'stand alone' instruction like NOP, requiring no extra data, then this phase would be the next instruction phase, identical to Step 1 above, except with the address bus now = 8001.)

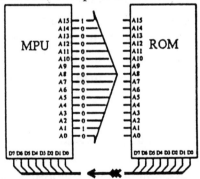

Figure 12.4 Step 4 The address bus is incremented to 8002 and the other byte of data is read into the processor.

Step 4

The address bus is incremented, to 8002, and another byte of data, required by the original instruction, is read in (Fig. 12.4).

179

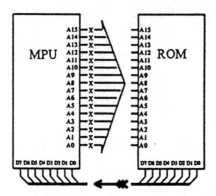

Figure 12.5 Step 5 The address bus might be incremented to 8003 or it may be changed to any address depending on the last instruction.

Step 5

The new address could be the next one up (8003), or it could be any address from 0000 to FFFF. It depends on the last instruction. For instance, if the last instruction was a JMP (jump) then the two data bytes that were read into the processor would have been the new address to jump to (two eight bit accesses are needed to complete a full sixteen bit address), and those two bytes would have been transferred straight on to the address bus (Fig. 12.5).

In assembler language the instruction would have looked something like this:

$$800 \text{ jmp XXXX}$$

and could have been assembled into memory rather as illustrated in Fig. 12.6. The operation is complete. An instruction has been fetched and executed. The processor will now automatically begin a fresh instruction fetch as it did at Step 1.

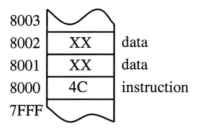

Figure 12.6

Crashing

After a reset, the processor will begin to run (execute) the program, performing alternate instruction and data fetches (or instruction fetches). It will always remain 'in sync' (synchronization) with the instructions.

What would happen if the processor got 'out of sync' and fetched data instead of an instruction during the instruction fetch cycle? The processor would execute that 'instruction' which might call for appended data, which might in turn be an instruction (or data)! In other words total chaos, hence the description crashing. The processing system would go 'off its head' doing completely unpredictable things.

A processor is totally reliable and will never crash of its own accord because its behaviour, although highly dynamic, is entirely 'mechanical', but the system in which it runs is vulnerable for various reasons:

1 Software: errors in the software (bugs).
2 System power supply: most modern systems have specialized hardware and supporting firmware to monitor incoming supply levels and react quickly enough to shut down the system in a controlled manner and restart it again correctly should the mains suffer 'brown out'. But some do not, in which case the drop in the

181

logic supply rail may be enough to crash the processor or crash the processor through unreliable memory operation, but not low enough to generate a reset, in which case it may stay crashed indefinitely or it may wreak havoc until it accidentally accesses its own reset vector or (if present and not regularly triggered by accident) a 'watchdog' times-out and resets the system (more on watchdogs below).

3 System unreliability: as the system ages it becomes particularly prone to 'mechanical' intermittencies in plugs and ic sockets (much more on this below).

4 System 'specifications exceeded': a system might work perfectly well at room temperature but not in the boot of a car on a hot day.

5 A fault: some permanent hardware faults can give very misleading symptoms, i.e. a gate's op might only drive to half its usual height (because it is damaged or loaded by another gate's faulty i/p). This damage is not enough for total failure. Indeed the system might run properly for hours, but occasionally crashes only to run again for hours after a reset. These problems sound impossible to tackle but this is not necessarily so. A 'heat gun' and 'freezer' might pin-point an ic in minutes by a worsening of the symptoms. A 'scoping of all the pins shows one of them with the wrong voltage levels. Note that this 'toughy' could be solved in minutes and without a diagram.

To recap, a processor has a tendency to increment its address bus. This is built into its 'architecture' so that it will always follow a program written sequentially, one byte after another from a particular low starting address, upwards.

The program itself can force the processor to make a jump or branch to any new address. This is part of the 'decision making' function that sets the processor apart from all other machines.

There is one last very important feature that must be understood by the fault finder and this is interrupt.

Interrupt

Processors work sequentially, doing one task after another, and often so rapidly that they appear to be able to do several things at once, but their operation is still sequential. If an input comes along and the processor is doing something else, it may not finish that task in time to deal with the input, e.g. keyboard input. If a key is tapped while the processor is busy accessing the disk drive or some other IO device it could miss the input from the keyboard. 'Unexpected' inputs like this are handled by interrupt.

Processors have one or more interrupt input pins which, when driven active, force the processor to stop whatever it is doing and respond. When the response is finished (and the active condition on the interrupt pin has cleared), the processor will restore itself and continue exactly what it was doing before it was interrupted.

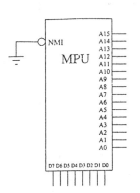

Figure 12.7

This response, called the interrupt routine, is written by the programmer but the interrupting action on the processor is built into the processor's design (architecture). This means that a working system could be totally 'hung up' by a fault condition that held an interrupt input active permanently (Fig. 12.7).

The non-maskable interrupt input is s/c to ground. If this input is level triggered then this processor will be forced to execute the interrupt routine for-ever, i.e. it will be permanently interrupted and unable to return to the main program. This would force the processor to continually try and service the interrupt forever, denying it the chance to recover to run the body of the program proper. The only way this fault could be diagnosed would be by carrying out the following tests.

a A test program showed interrupt always true (supposing that the test program can be loaded in the first place! Which may be possible if the fault is intermittent – load it in quickly while the going is good!)

b The fault finder deliberately looks for it, i.e. checks the interrupt pins are not 'jammed' permanently active (by some external fault).

A processor's interrupt pin(s) may be level triggered or edge triggered. An edge triggered input can still be 'held active' if, for instance, it is s/c to an address line, or clock, or data line, or if a partially damaged gate's op to it runs at half height. With the usual amount of cross-talk induced noise on it, the processor could detect a continual succession of edges (Fig. 12.8).

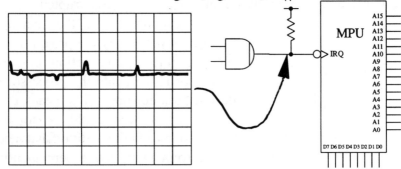

Figure 12.8 Even an edge detected interrupt input can be susceptible. A fault has left the input at about 2 V. The cross-talk induced noise renders the processor u/s.

Reset

Reset is the ultimate interrupt, that is, it dominates all others and therefore has the highest 'priority'. If a fault forces a permanent reset then the processor's address and data bus will be idle on all bits. These are identical symptoms to the state called no clock. If the reset duration is not long enough, entailing an unreliable reset, then occasionally the system may start and run perfectly indefinitely; otherwise it either won't start or it will start and crash later. In other words learn to look for the dominant condition:

a it does not matter how often or in what way a system crashes if
b it can be started and run indefinitely once.

No matter how infrequent the (b) case, it dominates in the sense that it 'stands out' (to the thoughtful fault finder) from all the 'colourful' symptoms of crash that surround it. For example:

1 Switch on – externally system appears 'dead'.
2 A subsequent cursory glance at the system buses (see Chapter 6) shows 'no faults'.
3 Switch off and on – CRTC programmed successfully (to come in Chapter 4) – but no keyboard activity.
4 Encouraged by this unexpected change, try switching off and on.
5 Dead again.
6 Switch off and on – keyboard responds – CRTC programmed – everything starts to work properly.

At this point the experienced fault finder will not switch off, instead all the functions of the machine should be tried out and then it should be left soaking and retried later. Suppose it continues to work well for the rest of the day. Now it is time to switch off and on again and the system is:

7 Dead again.

QED, it must be something to do with the reset circuitry.

System software
Software tends to be divided into 'operating system' software and 'application' or 'user' software. The operating system contains all the code used to start up the system after a reset and then to keep it

running, all hardware tasks like servicing interrupts or dealing with IO devices which interface with CRT controllers, disc drives, keyboards, etc.

'User' software could be an accounts program loaded into the system after it has booted up. When running, this software will need keyboard input and will also write VDU messages, etc. which are clearly operation system functions. The entire sequence would look like this:

a Switch on.
b Processor follows operating system code after reset.
c This code initializes processor: sets stack pointer, enables interrupts, etc.
d This code then initializes IO devices: to allow disc drives to work, VDU screens to appear (after programming CRTC) and keyboards to work, etc.

While the user program is run:

a Processor runs the program code which calls for keyboard input.
b Processor then runs operating system code to access keyboard through IO interface to discover the pressed key.
c Processor then returns to user program code with the key value.
d The program may decide to print that key on the VDU.
e Processor runs in the operating system code to access the VDU via the CRTC.
f Processor returns to the program code.

Try it out

Scope the address bus of a working system from A0 to the highest address line. Note the tendency for:

a More activity on lower order lines.
b How easy it is to inadvertently crash the system by accidentally short circuiting two adjacent lines together with the scope probe.

c If (and when) (b) happens note the total change in the appearance of the address bus (time base set lowish) from highly complex pattern (working) to very simple (crashed).

With practice a fault finder will get to know a particular system, perhaps favouring a particular data line as an instant check to see if the system is working properly.

Obs 'How do you know it's ok?'

FF 'Well, although it appears to be dead, from the lack of any IO activity, look at D7. Those faint tri-state curves there and that background 'ticking' effect once every second, don't you see?'

Obs 'Well, no I don't. You cannot tell what it is doing, not with a scope on a single data line!'

FF 'No of course I cannot tell what it is doing, but I don't care. The important thing is that this unit is doing exactly what working ones do!'

Obs 'But those patterns are far too complex for anybody to make a comparison.'

FF 'Exactly it is because the patterns are so complex that I can infer that they are identical.'

Obs 'How?'

FF 'Well, the highly complex pattern . . . there, between the recurring tri-state periods . . . it's more than 100 ms long meaning that the system is doing . . . 200 000 cycles or 50 000 instructions between "repeats" . . . that proves it is ok or very nearly ok. If it had any fundamental fault in the system buses it would be lucky to do more that ten instructions between repeats.'

Example

A 'control board' has just been rejected by the production line with the message 'will not start'.

The product in Fig. 12.9 has a processor running code from a single prom. You scope some of the lower order address lines and see these patterns, what is the fault?

Clue: Think about the production process and proms. (Also there will probably be seven more to come with the same fault).

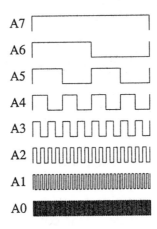

Figure 12.9

Answer: The prom is probably blank. Every location of a blank prom contains FF. The processor may treat this as no operation and will step automatically to the next instruction and so on. The address bus will therefore count continuously from 0000 to FFFF (the next location after FFFF is 0000), one address at a time without branching or jumping, and will look like the output of a binary counter.

Prom blowing is tedious. The operator loads the gang programmer with blank proms, eight (or sixteen or more) at a time and presses 'go'.

After weeks of this, the body takes over while the mind daydreams and sooner or later the operator is interrupted or disturbed by something and automatically removes the set of proms that have just been inserted without having programmed them. The programmer's display has the reassuring message 'verified' showing (from the previous set of proms) and the operator labels them up and off they go; blank.

Not all processors interpret FF as no operation, some will reject FF as an 'illegal' op code (instruction) forcing a reset action. In this case the patterns on the address and data buses will repeat after a few cycles.

13 Input/output devices

The IO (Input/Output) is the mechanism used by the processor to get 'data' to and from the outside world. This 'mechanism' is quite simple. Certain addresses are 'reserved' for 'IO devices'. Whenever these addresses are selected by the processor, ROM (or RAM) is deselected from the data bus and the IO device is connected instead and a byte of data can be written to it. This byte can then pass out of the IO device to a peripheral, e.g. a printer, during Output. If the IO cycle was Input then the reverse happens. Input from, e.g., a keyboard enters the IO device and then, when it is ready, the processor accesses the IO device and reads that byte. Before explaining this in more detail, it is necessary to understand how the data bus is accessed by more than one device.

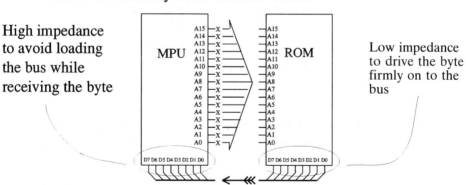

High impedance to avoid loading the bus while receiving the byte

Low impedance to drive the byte firmly on to the bus

Figure 13.1

To keep things simple, all the examples up to now have been read cycles (Fig. 13.1). The processor sets up a pattern on its address bus to select a unique byte in memory. As this is a read cycle, the processor will have set its data bus to input, i.e. to a high impedance 'listening state' and the memory will 'take over' by setting its data bus to output, i.e. to a low impedance 'talking state'.

During a write cycle, it would again set up a pattern on its address bus to elect a location in that memory but now the data would flow the opposite way, from processor to RAM (Fig. 13.2).

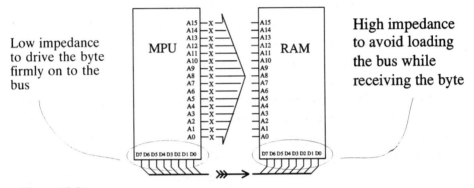

Low impedance to drive the byte firmly on to the bus

High impedance to avoid loading the bus while receiving the byte

Figure 13.2

To write the byte, the processor must turn on its data bus drivers to send all data bits high or low, i.e. it will drive the data bus hard with a low impedance drive. The memory must be in a listening condition, that is, all the memory's data bits must be high impedance to release the data bus to allow the processor to drive it, then the processor's byte will be able to flow from the processor to the memory.

Note that the address bus direction is always one way, from the processor to the memory, but the data bus can go in either direction, it is *bi-directional*.

The data bus is always driven either by the processor, during a write cycle, or by the memory, during a read cycle. We (the outside world) cannot drive the bus directly because we would clash with the processor or the memory. Similarly, if the processor tried to communicate with us, then it would unavoidably write to the memory as well. To avoid this 'clash of interests' we must create a space in the map, specially for IO, where memory is disabled (Fig. 13.3).

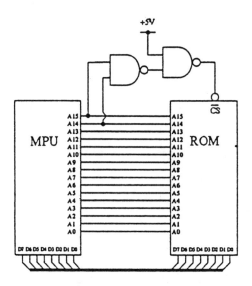

Figure 13.3 When the processor accesses any address between 0000 up to BFFF, then the memory will be chip selected. When both A15 and A14 are high (from C000 to FFFF) the memory will be deselected.

When the processor accesses any address with either A15 or A14 low, that is, any address between C000 up to BFFF, then the memory will be chip selected. When both A15 and A14 are high (from C000 to FFFF) the memory, via the lower NAND gate (which acts as an inverter), will be deselected (Fig. 13.4) and the memory map will appear as in Fig. 13.5, where the top quarter of the memory map will appear to the processor as an empty space.

193

Input/output devices

	A15 A14 A13 A12	A11 A10 A9 A8	A7 A6 A5 A4	A3 A2 A1 A0
FFFF	1 1 1 1	1 1 1 1	1 1 1 1	1 1 1 1
Exxx	1 1 1 0	X	X	X
Dxxx	1 1 0 1	X	X	X
C000	1 1 0 0	0 0 0 0	0 0 0 0	0 0 0 0

only A15 and A14 remain high	All the address lines from A13 to A0 switch between low and high as the address changes from C000 to FFFF

	A15 A14 A13 A12	A11 A10 A9 A8	A7 A6 A5 A4	A3 A2 A1 A0
BFFF	1 0 1 1	1 1 1 1	1 1 1 1	1 1 1 1
Axxx	1 0 1 0			
9xxx	1 0 0 1			
8xxx	1 0 0 0			
7xxx	0 1 1 1			
6xxx	0 1 1 0			
5xxx	0 1 0 1			
4xxx	0 1 0 0			
3xxx	0 0 1 1			
2xxx	0 0 1 0			
1xxx	0 0 0 1			
0000	0 0 0 0	0 0 0 0	0 0 0 0	0 0 0 0

All the address lines from A15 to A0 switch between low and high as the address changes from 0000 to BFFF.

But A15 and A14 never go high together

Figure 13.4

The ROM enable line, C000 to FFFF, has been decoded from the address bus by a NAND gate. The other NAND gate inverts it to C0000–FFFF bar which selects the ROM. An IO device could be added using the non-inverted C000–FFFF control to select it (Fig. 13.6).

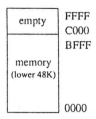

Figure 13.5

That device will then appear at every address from C000 to FFFF. This mechanism will allow the processor to access two different devices that share the same data bus. If the processor reads or writes any address between 0000 up to BFFF then the memory will be chip enabled allowing it to access the data bus and the IO device will not be chip enabled which will prevent it accessing the data bus. If the processor reads or writes any address in the range C000 to FFFF

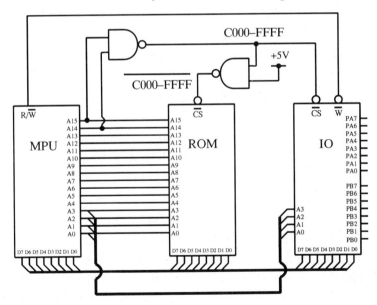

Figure 13.6

195

then the IO device will be chip enabled and the memory will be deselected. Without this switching mechanism these two devices could not share D7 to D0 because if the processor tried to read one of them, both would attempt to talk to the processor at the same time and they would clash.

We complete the diagram with four address lines A0, A1, A2, A3 and the bar write control to allow access to the IO's 16 internal registers. These 16 registers will now appear at C000, to C00F and repeat again at C010 to C01F and so on. In other words the 16 registers appear 1 k times over in the top 16 k of the memory map. If this is not clear do not worry too much, this type of knowledge is more the domain of the programmer but be aware of one important fact: if one of the address lines to the IO device is open circuit then that device cannot be programmed fully, which would make it appear faulty.

The main purpose of the IO device is very simple, it can be set up to pass a byte on the data bus to a port or it can be set up to pass a byte that has been placed on the port, to the data bus. The advantage is in the isolating of the eight bit ports from the eight bit bus with buffers. In the previous example of the last chapter where the LED was driven directly from decoded address lines, it was not possible to control it sensibly. With an IO device just about anything is possible provided that the item to be controlled is slower in operation than the processor controlling it.

Each bit of the IO port can be programmed not only as an input bit or as an output bit but also controlled independently, i.e. PA7 could be set up as input and PA6 could be set up as output.

Once a port has been programmed, for input or output, or for input and output on different bits, then subsequently the processor can write the output bits or read the input bits. Once the output bit(s) have been written, they will then stay that way indefinitely until the processor decides to rewrite them.

IO devices have other functions as well, like internal timers, which may be utilized by the system. Whether these internal features go faulty or not is less important to the fault finder. What is important is that they may drive an external output, e.g. irq (interrupt request), to interrupt the processor which, as we have seen, can have a devastating effect as a fault, but can be observed externally with a scope.

System is susceptible to IO devices

In fault finding, the most important thing about IO devices is that they are connected directly to the system bus. Anything sharing one or more data lines with the processor could crash the system.

In Fig. 13.7 the IO chip select is o/c. Therefore the IO device may 'turn on' or 'off' at any time, a disastrous thing! For example, whenever the processor attempts to read memory, there is the constant danger of the IO device butting in.

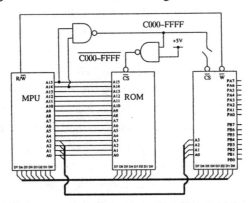

Figure 13.7 The IO device chip select is o/c. The IO is MOS technology, i.e. the CS input will float either high or low.

Should the CS input 'float' high during switch on then the system will initialize fairly well, the processor will be able to access the memory and anything else on the data bus, except of course the IO device. Should the chip enable drift low then there will be a catastrophic system failure because the IO will be selected

197

permanently for all the processor's address range and will clash with everything on the bus.

Try it

Find the IO device on your system and scope the chip select i/p(s). Switch on the system and notice the brief accesses to it once only during 'power up' to initialize the IO device. If there were any problem in this phase and the system could not access the device properly, then that device could 'hang' the system or the system might initialize properly at this stage but later accesses to peripherals connected through it might not work.

Keep the scope on that chip select and make the system use that IO by exercising a peripheral connected to it. Notice again the activity on that control line during the access.

If any other control lines, like A0, or bar write, or reset were not connected then the IO could again hang the system. The IO device is connected directly to the system's data bus or it may be buffered from it if the system has several data buses. This contact with a fundamental bus can be catastrophic if the IO device itself goes faulty. Again the IO device could hang the system.

If your system allows, disconnect the earth return(s) of the IO device, if that device is directly connected to the processor, then the system will hang. Examine the data bus and learn to recognize this pattern, it is typically caused by this sort of problem, clearly the IO device is 'dragging' all data bits towards V_{cc} as it tries to draw 'earth' currents through that bus. Scope the other controls connected to the IO device as well and notice that they are also loaded in a similar manner. When confronted with a problem like this on a complex system, we could put together all this information to help isolate the device:

a All data bits loaded.

b Bar write loaded.

c A0, A1, A2 and A3 are loaded but all other address bits are unaffected.

198

Conclusion: only a limited number of ics are connected to only A0, 1, 2 and 3 and bar write and eight bits of the data bus.

After initialization, the IO device will stay programmed that way until it is accessed again by the processor. One port might be an output (to a parallel printer) one port might be an input (from a keyboard). If we find that a port which is usually low impedance (output to printer) is floating, it may not necessarily be that the IO device is faulty, it might be that it has been wrongly programmed (a fault in the IO decode) or it might even be that a different program is running in the computer (a 'red herring').

The above example is a waste of address space. The IO device only requires sixteen addresses (A0 to A3). To reduce further the IO space in the address field, we must use more address lines.

Exercise

a Draw the memory map of the circuit in Fig. 13.8.
b In which direction does an irq go?
c If the processor is *writing* data, what state will A4 be in?

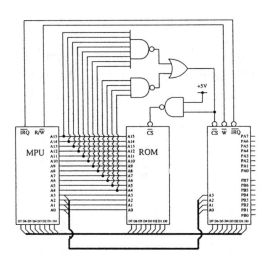

Figure 13.8

Answers:

a See Fig. 13.9.

b The irq is an input to the processor so the direction must be from IO to processor.

c There should never be any write cycles to ROM otherwise the processor and the ROM would clash: they would both talk at once. Therefore if there are any write cycles then the ROM must be disabled which can only occur during an IO access, i.e. when the address bus contains one of sixteen addresses between FFF0 and FFFF inclusive. For all of these addresses A4 is high.

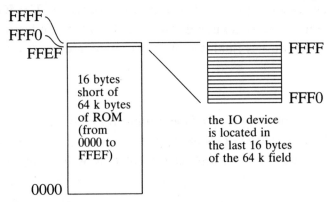

Figure 13.9

You may be wondering what the point of that last question was. Well, as a slightly premature example of what is to come, suppose we found the waveforms shown in Fig. 13.10 in the above circuit. What would it mean?

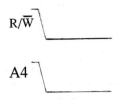

Figure 13.10

200

Answer: We know that no write should occur when A4 (or any address line from A4 to A15 inclusive) is (or are) low. So the waveforms in Fig. 13.10 mean that the processor is writing to ROM, an unpleasant experience for both processor and ROM!

There could be several reasons:

1 The wrong ROM has been fitted, i.e. wrong program.
2 The ROM is faulty.
3 If the system is working normally, and the ROM's contents subsequently prove to be in order, then it must be 'deliberate', i.e. a bug. (Use the other trace to discover which devices are contending and at what address(es) – see later chapters).
4 The system has crashed, i.e. we may have crashed it accidentally when scoping the system buses.
5 There is a fault in the IO decoding that deselects the ROM somewhere in its valid address range 0000 to FFEF when it should be selected. The effect of this would of course be devastating. One moment the processor would be happily fetching instructions from an accessible part of the ROM and the next moment (as the address changed into the 'faulty' area of the field) there would be no ROM! But the processor would still process, i.e. the high z state on its inputs would be read into the processor as valid data!

Do not worry too much if this is not entirely clear. It will become clearer if you come back to it having finished the later sections.

14 CRT control

To control a VDU (Visual Display Unit), a processor system must generate two signals:

a synchronization pulses;
b video.

The video is generated by specialized circuitry that converts a byte of data into a small portion of video. Video can be generated as a monochrome signal (black and white) or RGB (Red Green Blue). The processor has nothing to do with this process which is a purely automatic, slave-like activity. The processor will only initialize the video hardware at power up or perhaps when the display mode is changed, otherwise the video hardware constantly converts bytes of RAM into successive portions of the continuous video signal (Fig. 14.1).

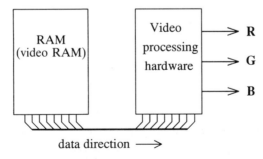

Figure 14.1 Converting data into video.

CRT control

The synchronization (or sync) pulses tie the addresses of the video memory to the face of the VDU screen. This is crucial, without a way to relate the video RAM to the VDU screen, each frame (of a stationary image) would show that image in different positions on the screen and it would be impossible to view.

The contents of the bytes of video RAM are always addressed and sent to the same positions on the face of the VDU at frame rate so that the image is 'locked' to the tube face (Fig. 14.2).

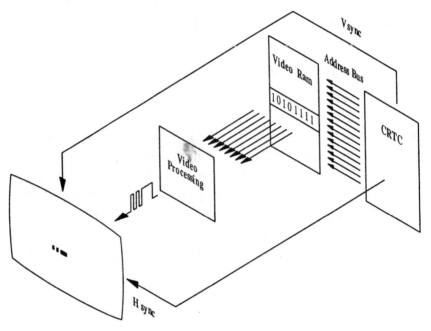

Figure 14.2

The CRTC addresses (Fig. 14.2) are rapidly scanning through the video RAM. As each byte is selected, its contents are rapidly processed and turned into a stream of video which controls the brightness (luminance) of the spot as it races across the face of the tube retracing each television line.

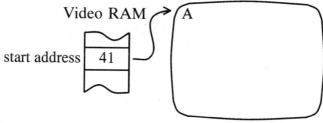

Figure 14.3

Every time a new VDU frame begins (top left hand corner of the VDU screen) the CRT controller always selects a special location. This is the video RAM start location. Anything placed in that location will always appear in the top left hand corner of the VDU (Fig. 14.3). (I have inserted the hex value 41 in the start location ASCII code for capital letter 'A' and indicated the position with that letter on the VDU although most displays are more complicated than this.) Now, because the VDU screen is being repeated at 50 Hz and because the video start address is always being read on to the screen at the same position each time, the character becomes visible.

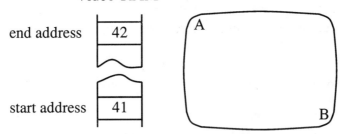

Figure 14.4

After beginning with the start address the next RAM address, video start address + 1, is selected and the next portion of the VDU is illuminated and so on until, just before the end of the frame, the last screen location and the last video RAM location are accessed (Fig. 14.4). Then a fresh frame is started by beginning the pass through the video RAM from the start address again and starting the new frame from the top left hand corner.

The video processing hardware has nothing to do with this synchronizing of video RAM and VDU, it only turns the '41' into a video stream that results in 'A' appearing. The synchronization is done by the CRT controller (Fig. 14.5).

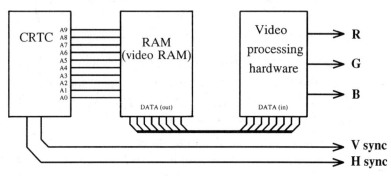

Figure 14.5

The VDU is controlled by HS (Horizontal Sync) and VS (Vertical Sync). Data to be displayed is selected by the CRTC address bus (nothing to do with the processor address bus). The CRTC controls both of these so that the same locations appear at the same positions on the face of the tube every time (Fig. 14.6).

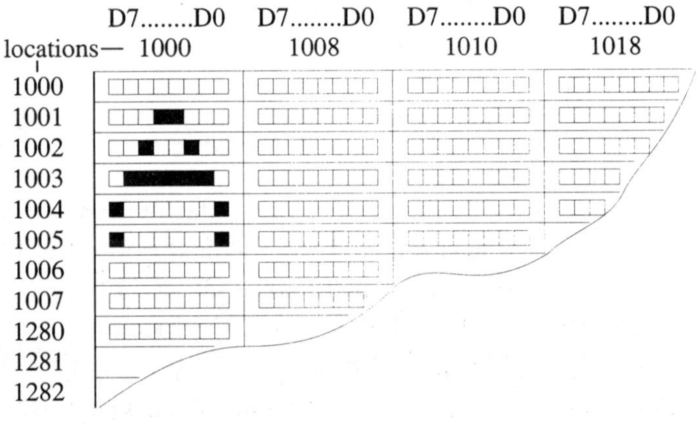

Figure 14.6

In other words, the screen looks like a window into the RAM. The bits of each RAM location might then represent 'black' with a nought and 'white' with a one. The processor would then be able to create characters, numbers or pictures by setting up appropriate 'bit patterns', e.g. the letter 'A' could be made by inserting 18, 24, 7E, 81 and 81 into address locations 1001, 1002, 1003, 1004 and 1005.

Note that there are all sorts of different screen formats to reproduce colour and luminance (grey scaling) for each 'bit' or 'pixel', e.g. if a system boasts 4096 different colours then evidently each pixel must have twelve bits of colour information associated with it. Anyway, this is not important to our purpose of understanding CRT control.

The processor also accesses video RAM

Although the processor and CRTC both access the same RAM, they are entirely independent. The CRTC plods away 'mechanically' counting through the video RAM addresses from start to finish fifty times a second. Its address bus would show that characteristic doubling effect of a binary count (Fig. 14.7).

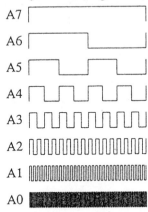

Figure 14.7 The characteristic doubling effect of a binary count. (Note that the pattern is not quite so simple. The 'row addresses' highest rate is one television line duration whereas the 'character address' highest rate is much faster (x no. of characters per line).)

CRT control

The 'slavish' nature of CRT control is evident on its address bus which endlessly counts through the same address range for each video frame, while the processor might access the video RAM at any time. Should the processor change the contents of location 1000 in the video RAM, when the CRTC might have just selected 1000 or might just be about to select 1001 or any address, it does not matter, they are independent and have no interest in the other's existence.

Evidently the address buses of the CRTC and processor must be multiplexed (switched alternately) into the video RAM otherwise they would clash with each other (Fig. 14.8). Note that the timing between the CRTC, MPU and multiplexer is carefully co-ordinated.

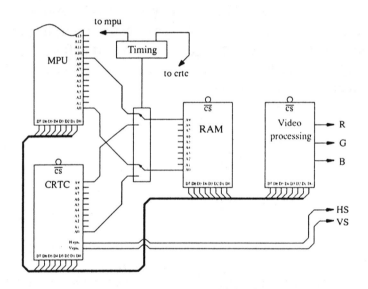

Figure 14.8

Why is the CRTC connected to the data bus?

Answer: The addresses selected by the CRTC and their positions on the VDU screen may change between different address 'modes'. A new mode might have more characters per line requiring more display RAM, in which case the CRTC address rate must increase to scan the extra address range in the same frame period. This feature is possible because CRT controllers can be programmed for different address rates. If a display mode change involves a change in video RAM requirements, then the processor will chip select the CRTC and reprogram it.

CRTC fault examples

In these examples the video is processed by a character generator. This device is economical with RAM in that it converts a single byte into an entire video character.

The start address of the video RAM is 0000 (say). If we placed 41, 42, 43, etc. into 0000, 0001, 0002, etc. then for the first frame's top left hand corner an 'A' would appear, followed by 'B', 'C', 'D', etc. (Fig. 14.9).

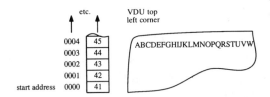

Figure 14.9

When the frame had finished the CRT controller would be forced back to the original starting address 0000 and would generate a fresh vertical sync to begin a fresh frame, and the above display would be repeated. As this happens at a field rate of 50 fields a

209

second, then our eyes (through the phenomenon of persistence of vision) would see the alphabet appearing on the VDU.

Once this has been set up the processor need not write to the RAM again. If the processor rewrites location 0000 with a new value, then the original letter 'A' will instantly vanish from the top left hand corner of the VDU to be replaced by something new, while the CRT controller chugs on regardless. Now suppose A0 from the CRTC were s/c to earth. What would happen?

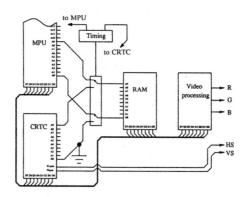

Figure 14.10 If A0 of the CRTC were s/c to earth at the input of the multiplexer, what would the symptoms be?

When the CRTC begins with address 0000 then all of its address lines will go low including A0. So location 0000 of the RAM will be accessed correctly and the result on the VDU will be as illustrated in Fig. 14.11.

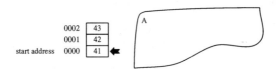

Figure 14.11

210

But when the CRTC counts the next location 0001, A0 will try to go high but, being s/c to ground, will stay low. The RAM will receive address 0000 again instead of 0001 so the contents of 0000, i.e. letter 'A', will appear on the face of the VDU where the letter 'B' appeared before (Fig. 14.12).

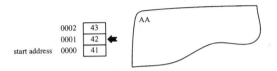

Figure 14.12

When the CRTC counts the next location 0002, then A0 will go low and A1 will go high, that is, the RAM will receive the correct address 0002 (for the wrong reason!) and will output the correct contents, letter 'C' (Fig. 14.13).

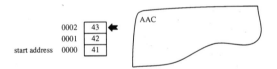

Figure 14.13

This error will be repeated for all the other CRTC addresses (Fig. 14.14), that is, each even address will be inaccessible (to the CRTC) and every odd address will be repeated.

Figure 14.14 Effect of CRTC A0 held low.

If the above fault had A0 tied high, instead of low, then the first address of the CRTC: 0000 would become 0001 at the RAM and instead of the RAM outputting 'A' it would output 'B'. The next address 0001 would appear at the RAM as the correct address (for the wrong reason) and so on (Fig. 14.15).

BBDDFFHHJJLLNNPPRRTTVVXX

Figure 14.15 Effect of CRTC A0 held high.

Note that these symptoms have a 'characteristic displacement of one position'. For higher order address lines the displacement effect is evidently greater, e.g. as in Figs 14.16, 14.17 and 14.18.

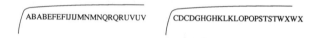

ABABEFEFIJIJMNMNQRQRUVUV   CDCDGHGHKLKLOPOPSTSTWXWX

Figure 14.16 A1 CRTC address errors (pairs are repeated).

ABCDABCDIJKLIJKLQRSTQRST   EFGHEFGHMNOPMNOPUVWXUVWX

Figure 14.17 A2 CRTC addressing errors (quads are repeated).

ABCDEFGHABCDEFGH   IJKLMNOPIJKLMNOP

Figure 14.18 A3 CRTC addressing errors (repeated sets of eight).

It can be seen therefore that for addressing errors, it is necessary to find the offset between repeats, and that the address line in error is then the index to the base of two that give that offset:

$$\text{offset} = 2^{(\text{address line})}$$

For example, if A12 were tied permanently low or high then:

$$4096 = 2^{12}$$

and large 4 k segments of the display would be displaced by 4 k locations.

Question: Would the above symptoms be reproduced by addressing errors on the MPU side, e.g. if the MPU's A0 were s/c to earth would similar addressing errors appear on the face of the VDU (Fig. 14.19)?

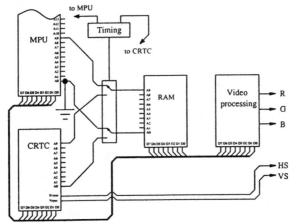

Figure 14.19 Would similar symptoms to those of the above addressing errors occur if A0 of the processor were s/c to earth?

Answer: With the processor's A0 s/c to earth the processor would crash!

If the fault lay in the multiplexer itself and did not load the processor's address line, the contents of the RAM would still be in

213

error, every other location that the processor accessed would not be inaccessible, half the data stored in the RAM would be lost and replaced with data repeated from wrong locations! If the address line were of a high order it may be possible for the processor to run successfully after a fashion because large pieces of memory would be contiguous (in one continuous unbroken block).

Row addresses errors

In the above fault examples the video is produced by a 'character generator', i.e. an entire video character is generated from a single byte. This is not as simple a process as it sounds bearing in mind that the character occupies several television lines at once, i.e. the character generator will have to transmit a series of television lines to build up the full character. But the CRTC will not be concerned about the individual lines of a character, which is controlled automatically by the character generator, instead the CRTC will only control the character positions with its M address lines.

Thus the CRTC's M address bus, labelled MA0 to MAx, is responsible for the actual character positions.

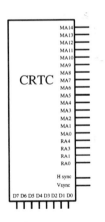

When the characters are 'bit mapped' to the screen, i.e. without the sophisticated television line processing of a character generator, then the individual television lines that make up a character must also be selected. The CRTC does this with its row address bus (RA0 to RAx). The CRTC address bus is split into two sections, MAx and RAx addresses. M addresses count whole character 'cells'. R addresses count the individual rows within a character cell (Fig. 14.20).

Figure 14.20

214

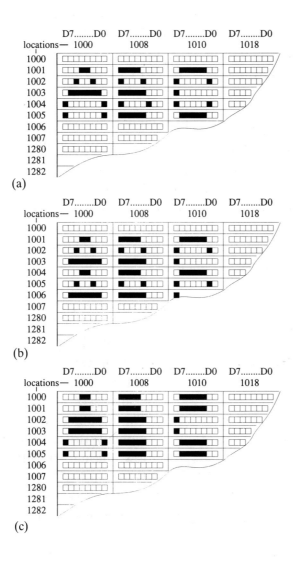

Figure 14.21

The row addresses select the row of the bit mapped character, while the M addresses 'select the entire character' so the above fault examples hold for bit mapped characters as well, i.e. an MA2 fault (repetitions in blocks of four) will be the same for either arrangement. But errors in the row address will produce a set of symptoms for the bit mapped characters, for example see Fig. 14.21(a)–(c). If the display at Fig. 14.21(a) is all right, what addressing errors could cause the faults shown in Figs. 14.21(b) and 14.21(c)?

Answers:
a RA2 tied permanently low at the input to the multiplexers.
b RA0 tied permanently high at the input to the multiplexers.

The faults in Figs. 14.21(b) and 14.21(c) are 'low z'. What would happen if an output driver failed high z, i.e. went o/c, leaving the following input floating? The chances are that the input would be 'driven' by nearby tracking through capacitive coupling and the display errors would 'flutter' between the address-line-tied-high and address-line-tied-low symptoms.

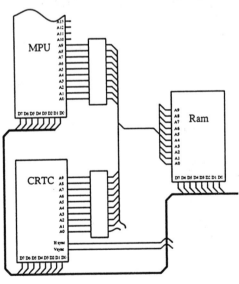

A multiplexer arrangement might exaggerate this type of fault, e.g. if an address line on the output side of the CRTC multiplexer was open circuit.

Fig. 14.22 illustrates an example of a 'high z' fault. The floating input will be violently driven either high or low during an MPU cycle. During a CRTC cycle, it will not be driven and so it will tend to stay at the previous level that it was driven to, during the previous MPU cycle.

Figure 14.22

216

During the CRTC cycle, the lower buffer will be enabled allowing the CRTC addresses to drive the RAM's address bus, and the upper buffer will be disabled, disconnecting the MPU's address bus.

During an MPU cycle the reverse happens, the MPU buffer will be enabled and the MPU's addresses will be connected to the RAM's address bus while the CRTC buffer will be disabled, disconnecting the CRTC address bus.

The fault (A0 on the output side of the CRTC buffer o/c) will prevent any drive to RAM's A0 during a CRTC cycle, but this line will be driven hard, either high or low, by the MPU during an MPU cycle. During the next CRTC cycle the line will not be driven and it will remain at the previous level that it was driven to by the MPU.

This will produce a 'fluttering' effect whenever the MPU accesses the RAM, e.g. if the above address line was MA0 then this display will appear:

CCmmuueess nnvvrr mmkk mmssaaee
o o p p t t rr   eeee   a a ee   1 1 t t kk ss

These two lines will appear superimposed and alternating rapidly (shown here separately). Although it is a little more difficult to identify, the recurring pairs displaced by two are the clue to a binary weight of A0.

If the above address line had been RA3 then the effect would have appeared something like this:

Computers never make mistakes
Computers never make mistakes
Computers never make mistakes
Computers never make mistakes

Again, the two lines will appear superimposed and alternating.

Programmable

The CRTC will be programmed once when the computer is switched on. If a system fault prevents any processing, then the CRTC will not be programmed. It will output the wrong count on its address lines and the frame and vertical sync outputs will run at the wrong frequency. This might be the first indication of a system fault.

If the CRTC is itself u/s or an IO decode circuitry fault prevents its being programmed properly, then the system may start successfully even though the screen might show typical symptoms of a 'dead' system.

Video interpretation faults

Faults in the video processing section, i.e. in the way details 'interpreted' as video, might appear similar to CRTC addressing faults at first glance, but they are quite different. This difference may only become apparent as we alter the display. For example, suppose D0 to a character generator was held low (Fig. 14.23), how would the display be affected?

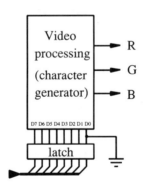

Figure 14.23

Answer: If we type 'B' hex code 42, then 'B' will appear. If we type 'C' (43) then another 'B' will appear because the lsb is held low. If we type 'D' (44) it will appear correctly. If we type 'E' (45) then 'D' will appear, etc. So the alphabet will appear as:

BBDDFFHHJJLLNNPPRRTTVVXX

apparently the same as an MA0 fault but not so: in this video interpretation fault, every key stroke produces a single character. In an MA0 fault every other key stroke produces two characters while the intermediate key strokes produce nothing.

That is, for all CRTC faults, half the RAM is inaccessible while the other half is accessed twice, each location becoming responsible for two VDU images.

In video interpretation faults, all the RAM locations are accessible but half of them are interpreted wrongly. The only way to find out which is which is to write to the screen and watch what happens.

15 Dynamic RAM (DRAM)

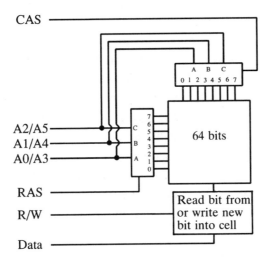

Figure 15.1 DRAM construction (contrived 64 bit DRAM)

The state of each bit is stored in a capacitor. These are selected by a single row and column. Rows and columns are decoded with address decoders whose inputs are combined together at the ic's pin-outs. To access a cell, first a row address (A0 to A2 in Fig. 15.1) is placed on the dynamic RAM's (DRAM's) address pins and then strobed into the row address decoder with RAS (row address strobe) where it is latched. The low order address is removed and replaced with the remaining high order bits (A3 to A5). This time a column address strobe (CAS) forces these addresses to the column address decoder.

The ic now has the full six bit address to select a single cell from the 64 cell matrix and this will either be output to the data out pin (if the bar write signal is high) or read into that cell (if the bar write signal is low).

As the cells are capacitors, they will gradually lose their charge and their state unless they are refreshed. This is done automatically whenever a row is addressed. All cells on the selected row have their states fed into individual 'sense' amplifiers that drive the decaying (or rising) voltage back to V_{cc} (or ground) restoring the cell's original state. Provided that all rows are selected within a given time, then all the cells of the DRAM will be refreshed whether or not any CAS cycles occur.

Note that a DRAM address pin will have two labels, the lower one will be correct for a RAS cycle while the upper label will be correct for a CAS cycle. A one mega bit DRAM will have ten address lines labelled: from A0/A10 to A9/A19 meaning, during the RAS cycle the ten address pins are A0 to A9 then later, during a CAS cycle, those same pins become A9 to A19.

Clearly, it is not possible to connect a processor's address bus directly to a DRAM. The processor's address bus must be split and multiplexed into the DRAM in two halves (Fig. 15.2).

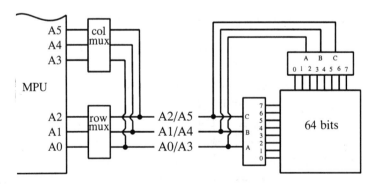

Figure 15.2

The processor would set up its address bus at the inputs of the multiplexers, and the lower half of the address only would be switched to the DRAM's address pins. Then a RAS would load that address into the DRAM, latched in the row address decoder. Next, the lower address multiplexer would be switched off and the upper address multiplexer switched on allowing the high order addresses through to the same DRAM pins but this time the DRAM would receive a CAS cycle, which would load the address into the column address decoder. Now the DRAM has the full address to select the cell.

Should another device need direct access to the memory, then it too would need its address multiplexed into the DRAMs in two halves, for instance the CRTC (Fig. 15.3).

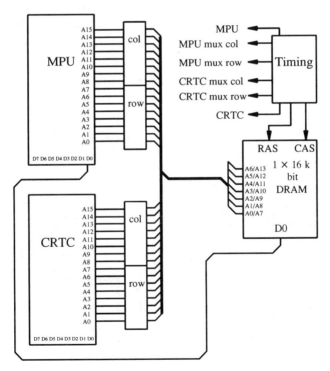

Figure 15.3

Dynamic RAM (DRAM)

Only one multiplexer can be switched on to the DRAM address bus at a time. Row address is always first. So the switching order must be from the bottom up, CRTC row, column, MPU row, column, then back to the CRTC again.

Note that the DRAM is only connected to D0, although it contains 16 384 bits, they are all D0. To make up the full eight bit data bus, another seven DRAMs are required (see Fig. 15.4).

This implies that for a RAM fault, only a single bit will be affected, that is, knowing the location of the faulty byte is no use, but knowing which bit is faulty will indicate which ic it is because each ic contains only one of the eight bits.

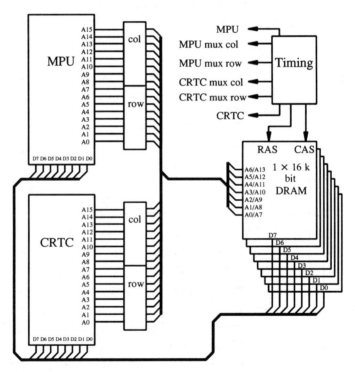

Figure 15.4

16 Digital systems fault finding

Ask a digital fault finding engineer how it is done and the reply might well be, 'Well, we check the buses and control lines' and then feeling that this sounds unsatisfactory adds, 'For any wrong signals.'

So what is a 'wrong signal' and how can a wrong signal help in tracing a cause? Before getting into that, let us talk a little more generally about failure types, namely:

a field failures;
b production failures.

Field failures are more 'genuine'. The system has worked well in the past and so the problem is more likely to be an electronic failure or mechanical deterioration (plugs and sockets).

Production failures can be almost anything, from component failure to wrong components to ridiculously unpredictable things like pcb tracking shorts between totally unrelated areas. But the production engineer has more resources and more specialized product knowledge and support than the field engineer, also the production engineer has not been up half the night driving to the site where the reception may not be convivial. Finally, the production engineer is blissfully unaware of the havoc that his bad handling of an ic socket is causing in the field two years later.

Mechanical connections are a menace, anything socketed or plugged is prone to failure; unfortunately, these shortcomings are seldom seen in production, and furthermore production engineers favour socketed connections – it makes life easier. In the field, however, in intermittent mechanical joint might crash some vast system. On arrival, the engineer will find everything working well. Even if by chance the crash is witnessed, fault finding will not be possible unless the fault remains permanently in place long enough to be found. If resetting the system cures the problem then it has gone and there is no point looking for it. If the problem can be reproduced, after resetting the system, by following a certain procedure, then this is not an intermittent problem, but rather a permanent problem with an 'intermittent' symptom. A truly intermittent problem has a certain unmistakable feel to it of a random 'mechanical' nature:

a Whenever the system is examined no fault is ever found.
b After an examination the system seems to run for longer than it usually does before crashing.
c After a time all the boards have been removed and taken back to base where they have been soaked and have all retested as functioning. (Soaked means left running for a while, e.g. a ten hour soak test means run for ten hours.)

The difficulty is that whenever the system is handled, the mechanical disturbance is 'fixing' the unreliable mechanical joint(s), for a time.

There is only one sensible course to take:

1 ease all socketed joints halfway out, then
2 flood the contacts with a cleaner like RS 556-648, and immediately push the joint back together. Warning, some plastics are vulnerable to this cleaner!

In pulling apart a joint, feel the contact pressure. Sometimes a joint falls apart because it has been previously mistreated (i.e. the socket must be replaced); sometimes the joint 'sticks', because flux 'wash' has got between the socket spring leaf and the housing and gone solid, restricting free spring movement. Some would condemn using contact cleaner as bad practice because it is a non-permanent repair.

This is true, and clearly if the customer can stand the cost, then replace all the sockets. Better still, remove the sockets and solder the ics in permanently.

If contact cleaner is used then it must not be the variety containing oil (switch cleaners) or the very volatile but gentle 'pcb cleaners'; use the aggressive stuff mentioned above from RS.

Finally, in defence of the 'quick fix', if the problem immediately reappears after the sockets have been disturbed and cleaned, then we know that it is a 'genuine' fault worthy of the excessive time it will take to find.

To return to the 'wrong signals' mentioned above. For fault finding we need an ordinary dual beam oscilloscope of 20 MHz bandwidth and 1 MΩ input impedance (or better) together with a times 10 probe. The probe, when set to times 10 must be trimmed to null the scope's input capacitance (see Chapter 17). Avoid using times one probes, they will damp out some of the subtle things we need to see.

System will not initialize

Recognizing whether a system has initialized or not is not always easy. If the system is a single eurocard mounted in a rack, the first indication, a failed serial link to it for instance, might be due to the customer selecting the wrong baud rate, for example, and nothing to do with the eurocard. For this very reason there is a tendency for manufacturers to build in an activity LED.

With small systems, learn to recognize the appearance of the system buses; there is usually a distinctive complex pattern particularly on the data bus. A production engineer will rapidly plug in one board after another with only a cursory glance at a particular data bit to distinguish nearly working examples from the non-initializers (with the objective of doing these first because they will be easier).

The larger the system, the more the evidence. If there is no display, no reaction to keyboard, or printer, activity LEDs will not blink, etc., then it is not initializing. If, in contrast, the screen start-up message or prompt appears, or any peripherals can be made to work, i.e. caps lock comes on when caps key pressed, then the system may not work fully but it certainly has initialized because a multitude of things must happen in the CPU (central processing unit) before IO devices can be programmed to set up peripherals properly.

In processor systems the buses and controls must be complete and intact before any (sensible) processing can begin. If any data lines are o/c, then the processor will not be able to execute more than a few instructions before crashing. If any address lines are o/c then the processor may crash immediately (for low order address lines) or it may run properly for longer (for high order address lines). If any data lines are s/c together or s/c to ground or V_{cc} then the system will crash almost immediately. It is likewise for the address bus, but again high order lines may have a less drastic effect. Any of the above failure conditions crop up in production failures. O/c tracks can be either 'etching faults' or dry joints. S/c can be 'etching faults' or solder 'whiskering' or bad handling (where an operative has peened one track against another) (or where an ic replacement has damaged adjacent tracking).

To check every track for continuity from node to node and then for an s/c to ground and then for an s/c to V_{cc} and then for an s/c between each other would be far too tedious. It is hard enough trying to follow a single track on a circuit board let alone 24 or 48, etc. – especially when, in production, only one failure in 50 might be of this type.

As an aid to physically locating the buses on a pcb, memorize the pinout of a 27xx series device (Fig. 16.1) and if fitted, use this as a starting point when dealing with an unfamiliar board. It is easy to memorize:

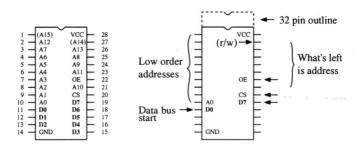

Figure 16.1

a The supply and ground are in top right and bottom left corners as they are on most logic ics.

b So from ground, the third pin up is D0, the rest of the data bus follows anticlockwise, like an ic pinout does, and continue counting up the right hand side to D7, then the next is the all important chip select so the other all important enabling control, output enable, is the next one up plus one.

c The low order address bus is mainly on the left hand side, except for the top two pins that are high order.

d High order lines are mixed up somewhere on the right hand side.

e The pin next to supply can be a high order address for the larger ROMs or it can be write bar for RAM.

f The same applies to 24 pin and 32 pin 27xx dil packages taking D0 third up from the bottom left as the starting point.

To return to the model in Fig. 16.1. All address and data lines are scoped quickly with the times 10 probe in a single pass. In doing

229

this, any 'irregularities' are noted mentally but not dwelt on, until all the lines have been scoped. This preamble is casual, just a quick $1/2$ second glance at each pin is sufficient. The idea is to get a 'feel' for the general behaviour. If your board has its system software in a 27 series device then you will be able to run the probe over the pins by touch without looking at the probe, so that you can concentrate on the scope's display. The purpose of this action is to reduce the boredom factor. One quick, simple action is all that is initially needed. From experience, you will be able to tell immediately that the system is working properly by recognizing particular patterns on the scope, or to tell that the system has a catastrophic error, like the possibilities in Fig. 16.2.

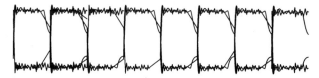

Figure 16.2

The data bus is all right. The other 7 bits are similarly complex. Note that the waveform is complicated further by the scope's triggering equally well on any of the 'cycles' (there are eight cycles in the diagram above), so that the scope cannot help displaying many different scans superimposed.

If these data were captured (for a processor running at 1 MHz,

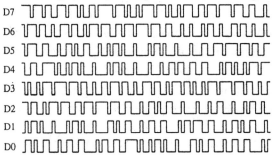

Figure 16.3 The sheer variety of patterns on a 'working' data bus will have a muddled appearance on a scope.

one second of data would be 250 000 bytes assuming an average of 4 cycles per instruction) for later examination, the thing we could be certain of would be variety (Fig. 16.3).

As already mentioned (see MPU action, Chapter 7) for 'catastrophic' failures, i.e. fundamental bussing faults, the bus patterns are almost always 'simple' (Fig. 16.4). This may not be so obvious in Fig. 16.4, but it is clearer on a scope trace, as in Fig. 16.5.

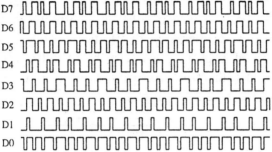

Figure 16.4 The often repetitive patterns in 'crash' will have a simple appearance on the scope.

Figure 16.5

The pattern repeats every few cycles. Note that this particular scope trace (Fig. 16.5) could equally well trigger from one of the

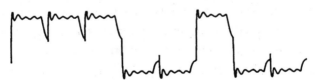

Figure 16.6

231

two positive edges: as Fig. 16.5 or as in the position shown in Fig. 16.6, which would not be possible to do if the data were 'valid', because it would not repeat itself so often, if at all.

The ringing at the end of each violent transition is the natural reaction of a bus's inherent capacitance and inductance. In addition, mixed in with this is 'cross-talk' picked up from digital activity in adjacent pcb tracking. This natural effect is useful for fault finding, but as it is only slight, we must take care not to dampen it with a 'times one' probe, always use 'times ten' correctly nulled.

Some designs deliberately damp the ringing effect with a resistive load tied to ground or V_{cc} (or both) to limit these undesirable excursions (otherwise excessive ringing may be seen by an input as extra logic pulses!) (Fig. 16.7).

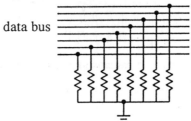

data bus

Figure 16.7 Data bus damping to limit ringing (and cross-talk). If a resistive element is o/c the fault symptoms might be rather subtle.

Bus damping

If this component were omitted, or one resistive element o/c, then the system may be unreliable, working most of the time, but occasionally failing. A production engineer, being accustomed to the one product, would notice the different appearance of the data bus: if the entire resistor pack was missing the whole bus would be more lively, or if only a single resistive element were missing (or o/c or DRT-joint) then only a single line would lack the loading effect, e.g. the resistor to D7 in Fig. 16.8 is u/s.

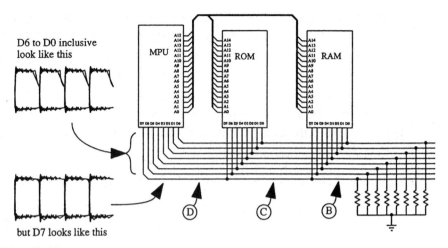

D6 to D0 inclusive
look like this

but D7 looks like this

Figure 16.8

Question: What else might cause this?

Answer: A break in the data line at B. Note that if there had been a break in the data line at C or D the effect, as we will see, would be quite different, but before we go any further, let us take a closer look at a 'cycle' (Fig. 16.9).

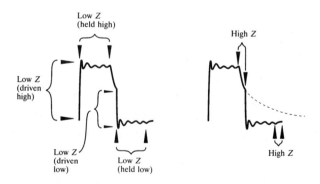

Figure 16.9

233

During this 'typical' cycle the data line goes tri-state (high Z) twice, just before the trailing edges. The dotted line is intended to highlight the effect of the resistor pack 'pulling' the high state back to ground (delayed by the natural stray capacitances of the bus and the devices connected to it (including the scope probe)). If the damping resistor is out of circuit then the data line will stay high during the upper tri-state period, held there by the stray capacitance, as in the D7 fault example in Fig. 16.8.

Question: If the cycle in Fig. 16.9 is a write cycle, then what is producing the low Z drive?

Answer: The processor. If the cycle was a read cycle then either the ROM or the RAM (or an IO device) would be driving the data bus (sending the byte back to the processor).

TTL input loading

TTL inputs will try to pull a floating bus to 2 V (Fig. 16.10). In contrast to this, NMOS and CMOS inputs have no pulling effect, so, what could the fault in Fig. 16.11 be?

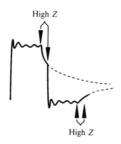

Figure 16.10 The loading effect of a TTL input will pull any tri-state periods towards a 2 V level.

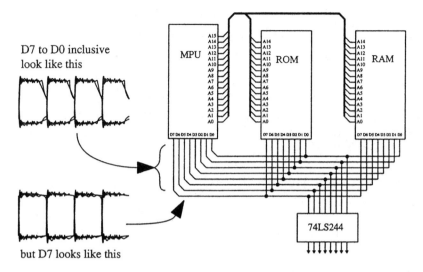

D7 to D0 inclusive
look like this

but D7 looks like this

Figure 16.11

Answer: The processor and the memories will be either NMOS or CMOS. The buffer is evidently TTL and the patterns on D6 to D0 confirm this. But there is no TTL loading effect on D7, so it cannot be connected to the buffer, i.e. there is either a 'mechanical' break in the tracking or the buffer itself is faulty (o/c) on its D7 input. The 'complex' data bus patterns, including D7, indicate that the processor is probably working properly which implies that the address and data buses are complete and intact between the processor, the ROM and the RAM. Therefore the track break, if there is one, must lie somewhere between the input to the buffer and the node where it connects with the data bus otherwise the buffer is u/s on its D7 input.

In a wholly CMOS system buses will be driven symmetrically between 0 and 5 V by the symmetrical CMOS output. In TTL and NMOS systems, buses will typically be driven low harder than they are pulled high. When the two technologies are mixed a trace like the one illustrated in Fig. 16.12 is produced.

235

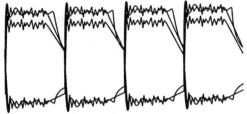

Figure 16.12 Mixed technologies on the same bus: CMOS, NMOS and TTL.

If the system in Fig. 16.12 usually looked like that on all data bits, but became like Fig. 16.13 in fault, then it would appear that the single NMOS device had failed catastrophically on its data bus output.

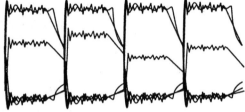

Figure 16.13 Fig. 16.12 bus in fault.

Obs 'You just said that a catastrophic failure would give simple data bus patterns.'

FF 'Yes.'

Obs 'Well the above failure looks complex.'

FF 'Yes, so therefore the processor, ROM and RAM and all the interconnects are almost certainly ok.'

Obs 'So the NMOS device is probably not one of these?'

FF 'Exactly.'

Obs 'It all seems rather fortuitous. What if this fault had happened in a simple system with only a few devices on the bus, and no "pull down" effect?'

FF 'True, but then there would be less devices to choose from. The more complex the system the more the clues. If the faulty device's output had gone permanently high Z the "touch test" (see page 169) would show up those periods, e.g. (see Fig. 16.14) in a simple system, any high Z states are not immediately obvious (Fig. 16.14(a)) until a finger is dabbed on (Fig. 16.14(b)).

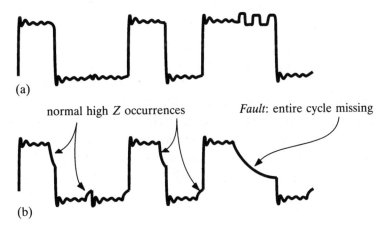

Figure 16.14

Note that there is a subtle clue in the upper trace: there is more 'pickup' on the data line during the missing cycle.

Contention

If two lines are s/c together or if two bus drivers are enabled (wrongly) together, then those two lines or that whole bus will show signs of contention as illustrated in Fig. 16.15.

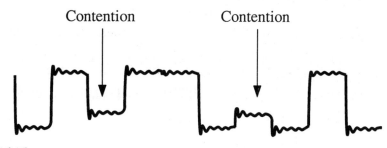

Figure 16.15

237

During the first cycle both lines go low together. During the second cycle both lines go high together but during the third cycle one line goes high and the other goes low and they clash producing a level that is in between. In Fig. 16.15 the level is below midway (1.5 V) which is typical of contention between NMOS drivers. If two CMOS output drivers clashed, the level would be midway (2.5 V).

Note that throughout the above trace, all levels are low Z. Sometimes contending cycles and floating cycles appear to be similar but contention will not be affected by the finger test.

Some microcontrollers (microcomputers-on-a-chip or custom gate arrays) have open collector outputs to the 'outside world' with very weak pull ups (50 kΩ to 100 kΩ?). Contention between these pins will evidently not show; because the active pull down will win, grounding the level.

The hard part

Having found that a particular line is s/c to another is one thing: finding the short itself (without disfiguring the pcb) is quite another. A 'Tone Ohm' (emits an audible tone that changes with resistance) or 4½ digit dvm can pinpoint it, if we are able to follow the tracking (a 'bare board' is very useful, if not essential).

A strong light under the pcb can reveal tiny 'whiskery hair line etching' shorts (hair or dust contamination during the photo/etch resist process), but not always because legends, components or the earth/supply planes of multi-layer boards block the light.

If the s/c is invisible, it could be 'track blasted' with a high current. Set the power supply to 0.2 V or lower and gradually increase the current (by reducing the current limit) across the s/c, obviously making the local connections as close as possible to the s/c. This may be frowned on as bad practice for two reasons:

a Pcb tracks can be destroyed.

b The engineer might forget to reduce the voltage to 0.2 V (if he uses the same psu for powering the uut, then many components can be destroyed (best case) to partially damaged (worst case).

All of us will sooner or later commit (a) and (b) but the alternative can be scrapping an entire pcb (for an s/c to an internal layer in a multi-layer pcb) or accidental damage when scraping away at the pcb surface with a scalpel (the favoured method for removing copper). (Evidently if a current of 5 A will not blow the s/c away then it must be big enough to see.) It all depends on the work environment; a cheap consumer product is generally relaxed while Ministry of Defence work will be very strict.

Using the scope to locate short circuits

A safe, quick way to isolate the s/c on a contending bus is with the scope. During the interval of collision there must inevitably be some ringing, this will be larger at the driven ends than at the shorted node (Fig. 16.16). Alternatively, watch the scope image carefully and deliberately slide the probe's tip across the observed pin until it touches the adjacent pin that it is shorted to (Fig. 16.17). Do this alternately and there will be a slight change in the scope's trace due to the ringing during the contention, evidently the closer we get to the actual short, the less the change, which would appear to be nearer to the device on the left (the lower trace is probe's s/c on).

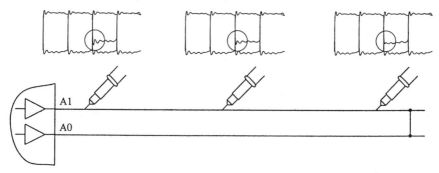

Figure 16.16

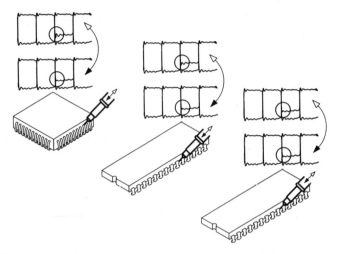

Figure 16.17

Try it out

Try blobbing some solder across pins 10 and 9 of the 27xx device
(A0 s/c to A1). Scope A0 at the processor or at another 27xx device.
Deliberately s/c A0 to A1 with a sliding action so that the probe
never leaves the A0 pin. Do that repeatedly and watch for a tiny
change on the scope as the short is taken on and off. Repeat this
action at the pins of the ic which has the soldered s/c, here there will
be no change.

Output s/c to earth

The same thing applies if an output is s/c to ground. At the output a
small 'differentiated' signal appears instead of the square 2 MHz
wave. By 'walking' the earth of the scope and the probe towards the
'terminated' end, i.e. where the s/c is, this tiny waveform
diminishes.

240

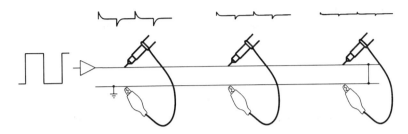

Figure 16.18

Obs 'Hang on a minute, what's wrong with simply using the DMM to locate the short?'

FF 'Nothing, if the DMM is sensitive enough, but mine isn't and besides, I have to power down before I can use the meter.'

Obs 'Ah ha, laziness!'

FF 'No; efficiency, but there is another possibility. If one of many common bus drivers goes permanently low Z then this method might pinpoint the device.'

Obs 'What about the chase-the-chip-select method?'

FF 'No, I mean permanently low Z, i.e. irrespective of chip select. The meter would be useless because the fault would only show when powered up.'

Obs 'But suppose one device was permanently driving the bus, then it would collide with every other bus driver sooner or later, so whichever chip's output you selected to view would also be low Z during the collision and therefore exhibit damped ringing.'

FF 'But only one of those devices would be in contention during all collisions.'

Obs 'Ah, so only one device would appear "quieter" all the time. Sounds a bit dubious to me, have you tried it?'

FF 'No, not yet.'

Obs 'I've got a better idea, why not use a current tracer?'

FF 'I would if there was one available!'

Obs 'Why is it that you never seem to have any decent equipment? Even that scope you're using is an old linear 20 MHz specimen.'

FF 'Because that's the way it is in industry, but I'm not complaining. I prefer simple fault finding equipment, the less sophisticated, the better. There's less to go wrong, less to learn and to remember.'

Tri-state cycles on the data bus

If one data line is o/c then whole processor cycles will be 'missing', that is, the line will float tri-state during a read or a write cycle. Depending on the 'passive' loads on that bus, this tri-state period may show as an exponential curve, otherwise it may be necessary to touch it with a finger to prove the point. As some cycles are missing, then that data line will appear to be less active or 'running slower' than the other data lines.

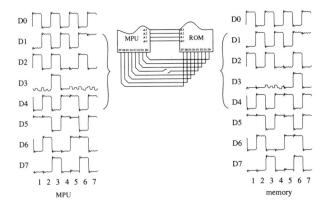

Figure 16.19

Fig. 16.19 shows scope traces of the above fault at the MPU end of the data bus and scope traces at the memory end. Naturally the same data lines should be the same, but because D3 is o/c, then either end of D3 is different. Each of the cycles is numbered.

242

Question: Look at D3 during cycle number three and cycle six. Which cycle is a read cycle and which is a write cycle?

Answer: Cycle three is a write cycle because, at the processor pins, this line is driven. The memory cannot be driving it here because the line is o/c from the memory so it must be driven by the processor. If the processor is driving the data bus then it is writing, so it must be a write cycle. Similarly, during cycle number six the line is driven, but this time it is only driven at the memory end, if the memory is driving the data bus then the memory is being read, so it must be a read cycle.

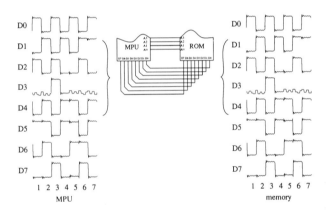

Figure 16.20 Tri-state is on data bus (D3), the data bus itself is intact.

In Fig. 16.20, the data bus is intact and so D3 looks identical at either end of the data bus. There are large tri-state periods and the frequency of D3 is slower than all the other data bits, so clearly either the processor is damaged on D3 and cannot drive that bit or the memory is damaged on D3 and is not driving that bit, but which one is it?

To find out, compare the r/w (read/write) line with the faulty bit (Fig. 16.21).

243

Digital systems fault finding

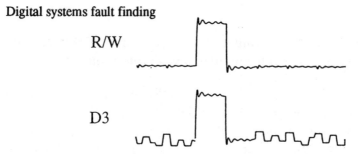

R/W

D3

Figure 16.21 Which D3 driver is damaged, the MPU or the memory?

Answer: The processor is faulty because the tri-state periods coincide with the write cycles, i.e. during a processor write cycle, D0, D1, D2, D4, D5, D6 and D7 all drive their lines but D3 is undriven.

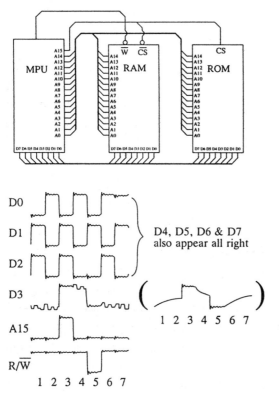

D4, D5, D6 & D7 also appear all right

Figure 16.22

What is the fault in Fig. 16.22? (The waveform in brackets shows how D3 is affected by the 'finger test'.)

Answer: The RAM bit 3 output driver is o/c or the D3 pcb tracking to the RAM is o/c. During a write cycle D3 is driven without any problem. During read cycles D3 is tri-state on some cycles. But D3 is driven during one of the read cycles, the third one when A15 goes high, i.e. when the ROM is selected (A15 high selects the ROM, the symbol on the ROM chip select indicates active positive), so the ROM is all right on bit 3, but all the other read cycles, where D3 is floating, happen while A15 is low. When A15 is low the RAM is selected (the chip select has a negation symbol) therefore the RAM is not driving the bus on D3.

Address bus 'dominates' data bus

The address bus is simpler than the data bus because it only goes one way. If an address line is o/c then it will always appear stationary (loaded to ground via the scope).

If a fault developed in the address bus, i.e. if one more data line were held permanently high, low or at some midway position or if two lines were s/c together or if a line was s/c to any other digital (or low *Z* analogue) signal, then evidently the processor would crash.

Now, in a crashed condition that the processor can never recover from because the address bus is at fault, the processor could do anything, e.g. it might:

1 read addresses that contain nothing or are write only devices which would produce tri-state on the data bus;
2 write to read only devices producing contention on the data bus.

Note that in both of these cases the tri-state and contention will appear across all data bits. In other words fundamental problems might exist in the address bus which give rise to catastrophic results on the data bus. The reverse of this cannot happen, so the rule must be: If errors show in the address and data buses, always sort out the address bus error first.

Static address line

If an address line appeared permanently in one state, it could be s/c or loaded to ground or supply, or it could be that the processor is currently stuck in a tight loop running 'code' in a limited address range. This range may not change the address line under inspection especially for higher order address lines. The lower order address lines will be inherently more active. A0 for instance will never appear stationary, because it must change state every time the address changes from odd to even. A15 by contrast will only change state once as the processor crosses the middle boundary of the 64 k address range, from 7FFF to 8000 or from 8000 to 7FFF. Any address, data or control bus fault will crash the processor and as a result some of the higher order lines might appear stationary. If they do appear stationary then we cannot tell whether they are stationary because they are in fault or because some other fault has crashed the system.

A quick method to force the processor to drive these address lines, especially the higher orders, irrespective of any fault conditions, is to reset it. So the procedure is to reset the system as each stationary line is scoped. All address lines will almost certainly leap up and down at least once, proving that they are not loaded to anything. It may take a little experience to notice these faint 'transients', it may be necessary to turn the brightness up, or occasionally to reduce the time base to minimum to see a single 1 microsecond pulse, but that is all that is required for proof.

A simple way to initiate a reset, without interrupting the power supply, could be to s/c the output leg of a 7805 series regulator with the adjacent centre leg (earth) with a screwdriver. All these devices are current limited and cannot be damaged. Otherwise we might have to examine the reset circuitry for some safe way to initiate it. Evidently if one address line does appear stationary despite a reset, and if everything else looks all right, then it is worth further investigation, but before turning off the power, just try temporarily shorting that line to another active line with the tip of the scope, if the active line is completely unaffected by the imposed load of the 'dead' line which means that the 'dead' line accepts the drive from the active line without loading it, then clearly the 'dead' line is floating tri-state which must never be, so either its path to the processor is broken, or the processor is u/s on that line. If, in contrast, the 'dead' line loads the active line absolutely so that the active line is totally overcome and adopts the 'dead' line's state, then clearly the 'dead' line is permanently low impedance and might be s/c to earth or supply.

Finally a third possibility is that both lines take on the active state, but there are periods of contention. This means that the processor was active on that address line after all but the activity was not noticed.

If two address lines are s/c to each other, then we could use the scope to isolate the s/c, with the above method. When data or address lines are s/c to each other they will tend to be adjacent because the pcb layout must be adjacent.

If we are unlucky enough to have a s/c between a bus line and something else, then it may be possible to reason what that something else might be. Has it a characteristic frequency? If so, where in the circuit is this frequency generated, etc.?

What to do next

All data and address lines look to be working properly. This means that:

a   The patterns are complex in the typical way of a working system with no contention or tri-state periods occupying a full machine cycle (or more).

b   The patterns look the same as other identical systems that work all right. There are tri-state periods and even a little contention but as these appear identically in the working systems they are not indicative of fault. Some possibilities are:

(i) software problem;
(ii) memory fault;
(iii) a 'timing' problem;
(iv) a faulty processor or support ic.

In (i) an earlier and now redundant version of the operating system software has been inadvertently blown or a corrupt version of the correct software has been blown into a prom. If this is the case then the problem will tend to be in multiples of eight or whatever quantity the gang prommer uses.

In (ii) either the prom was blown correctly with the correct version and subsequently its contents changed – very unlikely – or the RAM nearly works but has a small portion of it unreliable – possible. It is possible for dynamic RAM to have no refresh at all but still 'work' (where a processor 'background' activity keeps the DRAMs alive). A wrong memory part might have been fitted. Either the wrong size, or perhaps the correct size but wrong access time.

In (iii) 'timing problem' means a subtle change in any clock, control or bus characteristics anywhere. If the supply to the system is slightly high or low then a 'timing' problem might occur. If the system is connected via a communication link to another system then a small change in the main clock frequency could change the link communication speed enough to make it unreliable or u/s.

If attempts to find reasons for (i), (ii) and (iii) have failed, then (iv) must be assumed. There may be no choice but trial and error replacement, an undesirable but sometimes necessary chore.

If a particular hard-to-find fault recurs, then there may be no fault as such but a design error. The few regular drop outs might be the tip of the iceberg. If a product is proven and then suddenly a run of similar problems occur, then the easiest approach might be to look for a change in ic supplier, or perhaps batch code numbers between an older build sample and the current build. If replacing the new RAM type with the old fixes the problem, it may not be a faulty batch but a bad design which the old RAM's superior specification obscured.

Heating the pcb with a heat gun or cooling it by freezing or raising or lowering V_{cc} slightly could make the system burst into life. Careful localized heating or cooling would then point out the faulty area. Before any fault finding effort is wasted, it is best to verify first that a working product cannot also be made to fail by these methods. If so, then again there may be a general design flaw and not a fault.

Supply s/c to ground

Use voltage measurement rather than resistance to pinpoint the short, i.e. the average DMM which can resolve 0.1 Ω (not good enough for this type of problem) will also resolve 1 mV. At 1 A, that gives 100 mV drop across 0.1 Ω, i.e. a 100 times the resolution of the ohm range. If the onboard regulator will not deliver sufficient short circuit current (before thermal shutdown) then bypass it (Fig. 16.23).

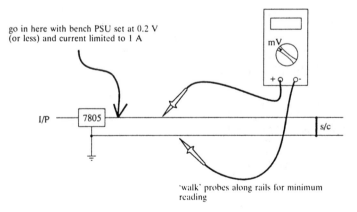

go in here with bench PSU set at 0.2 V (or less) and current limited to 1 A

mV

+ Q Q-

I/P — 7805

s/c

'walk' probes along rails for minimum reading

Figure 16.23

Wind the output of the psu down to 0.2 V (to avoid damage to the onboard regulator should the s/c 'vaporize') and increase the current limit to (say) an amp and 'walk' the probes along the rails to locate the s/c. Evidently, at one amp, every mV is equivalent to 0.001 Ω.

Warning

If the s/c is inaccessible, e.g. when it is between the inner layers of a multi-layer board, it may be possible to 'track blast' it with high currents. But watch out – the following cases can have disastrous results:

a The (internal) 12 V plane is s/c to the (internal) 0 V plane, and the fault finder winds the current limit up to the maximum of 10A to try and blast it away.

b It is only drawing 5A, it appears to be a very tiny short (to have a resistance of 2 Ω) and therefore possible to blast. Meanwhile the current starts to fall off slowly towards 4 A.

c It begins to get hot – maybe melting - and it looks as though a little more current and it will go; the fault finder winds up the supply voltage knob (evidently the meter hardly moves from 0 V). The current increases gradually to 6 A, 7 A then falls back a bit, but increases again as the 'voltage is turned up', and suddenly the needle falls back to zero as the s/c is vaporized.

Now the supply voltage jumps from zero to more than 20 V and all the electrolytics explode!

Serial IO

The unit in Fig. 16.24 works intermittently on the serial port. The test engineer starts to examine it with a scope.

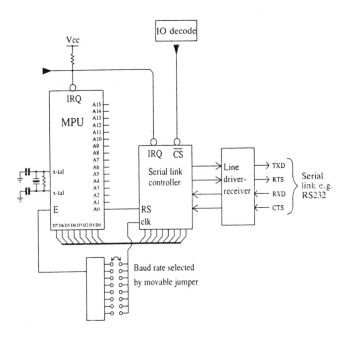

Figure 16.24

FF 'TXD is . . . minus 9 V and . . . so is RTS, that's ok.'

Obs 'How do you know what the levels should be?'

FF 'Because when switched off and on, i.e. reset, then both RTS and TXD are always programmed low.'

Obs 'Does this apply to all RS-232 equipment?'

FF 'No idea, even if it did, I could never remember it! So I just compare the faulty one with whatever is available.'

Obs 'Can't you just ask engineering what the states should be?'

FF 'Yes but the person who wrote the driver firmware will never remember either, anyway let's get on, we only have fifteen minutes maximum or the board is scrap.'

Obs 'If this unit works sometimes and not others, then surely that will have nothing to do with the output (or input) levels? I mean it's digital isn't it, so surely it will either work or it won't.'

251

FF 'It should either work all the time or not at all but occasionally, very occasionally, a digital level might, due to damage to a driver or receiver, be running between wrong level(s) so that it only just exceeds an input threshold. Then some tiny external influence like temperature or pickup may be enough to push it over the edge – now it works – now it doesn't.

Obs 'I think I see but I find it hard to believe!'

The FF links the RS-232 port to a terminal and attempts to get the link up and running. Suddenly it starts working and he scopes the four lines alternately.

FF 'Look, while it is working I quickly scope all four lines RTS, CTS, TXD and RXD and they are all running correctly between plus and minus 9 V.'

The FF then continues scoping the four lines alternately while heating the pcb with the heat gun and handling it roughly. Then he tries freezer.

FF 'Look it's started to fail, notice that the four lines are, at the same instant, still running at the correct heights.'

Obs 'Proving that this fault has nothing to do with wrong height levels!'

FF 'Exactly.'

Obs 'The freezer seemed to make it worse.'

FF 'There is something much more important than that. During the fault, the activity on the four lines continued.'

Obs 'Meaning what?'

FF 'Meaning that the fault it not a "catastrophic" type of fault – like failure of the serial controller – it is more of a subtle type failure, as the symptoms suggest.'

Obs 'Oh I see, if it had turned out that a faulty driver/receiver had output/input wrong levels then that would be a permanent fault with intermittent symptoms.'

FF 'And this, so far, seems to be an intermittent fault.'

Obs 'That's why you were roughly handling the board, to see if the intermittency is "mechanical"?'

FF 'Yes, but it could still be a permanent fault. The interrupt line could be running at 2 V, e.g. if the pull resistor was missing or DRT-joint, then noise on that extensive-shared-line could trigger unwanted interrupts now and again, for instance condensation from the freezer might increase this pickup past the thresholds bringing on the fault.'

Obs 'But all you have to do is scope it.'

The FF scopes irq at the processor's pin and at the serial controller's pin and finds good 0–5 V logic levels while the fault is active.

Obs 'That's a shame. Maybe it is the serial controller after all, perhaps it has some subtle internal fault, or perhaps the processor does not always respond to irq?'

FF 'Possible but unlikely, processors and other large devices are rarely at fault (except for IO devices).'

Obs 'You are now trying the freezer and heat gun alternately, to pinpoint the faulty area.'

FF 'Yes, but look at this, if I put my hand here, it works, if I move my hand away, it stops.'

Obs 'It's the . . . crystal, the crystal is faulty?'

FF 'No, if the crystal stopped then the entire system would crash.'

Obs 'No, I mean the crystal is being pulled off frequency by your hand so that the baud rate for the RS-232 goes out of spec' and the link then fails.'

FF 'Nearly correct, look one of the 22p capacitors is missing.'

Obs 'Well that is amazing, what a hard fault!'

FF 'It is, but then it isn't. It's hard because there are so few things to go wrong and no logical fault finding path to follow. But then again because there are so few bits it's easy, I mean, the chances of it's being a device fault (apart from the driver/ receiver) are slim, so it's not the processor or the serial controller, then it must be either:

a "partial damage" in the driver/receiver or what is more likely an error in the "termination" components connected across the RS-232 lines;
b intermittent chip selection via the IO decode;
c intermittent interrupt request;
d "intermittent" clock.

All these can be checked out quickly with a scope except for (d) which needs a frequency counter which we do not have readily to hand.'

Obs 'But why couldn't it have been an intermittent socketed joint or dry joint at the serial controller pins?'

FF 'Why didn't you say processor's pins?'

Obs 'Because the entire system would have crashed.'

FF 'Well the same could apply to the serial controller, if the earth return went o/c, but the main reason is that the serial controller is programmable so that if a data or control line went o/c briefly the symptoms may have been more lasting (because of an accidental reprogramming that may not be corrected until the unit is switched off and on).'

Microcontrollers

A microprocessor is useless without memory and IO. For any processor system there is always this fixed additional overhead.

A microcontroller combines all three functions in a single package, i.e. an ideal design will have no external ROM or RAM, no external serial controller, no external IO devices or latches. Instead there will be a single microcontroller and a crystal to drive it. These will sit at the centre of whatever it is they are controlling with the pins of the controller radiating out to the surrounding circuitry which need not be entirely digital.

Most pins of the controller may be IO but some may be analogue, e.g. AD input. Some IO pins may double up as the original address, data and control lines of the old microprocessor because there is often an option to use a controller as a processor. The usual reason for this is the limited size of the controller's built-in memory.

After a reset, the controller will need to know whether it is supposed to refer to its own internal memory (and subsequently have most of its external pins free to use for IO) or whether to access memory externally by pre-configuring most of its pins as ordinary address, data and control lines. It does this by reading the input 'code' on its 'mode' pins, just after a reset. These pins may double up as general IO, in which case they will be pulled high or low gently by largish resistors so that, should they revert to output, they will easily overcome these 'weak' pullups/-downs.

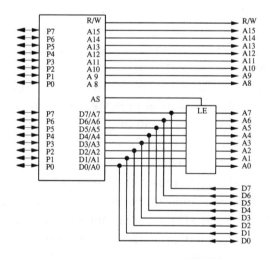

Figure 16.25 Microcontroller used in extended addressing mode. Two of its IO ports and two control lines, AS and R/W, allow direct connection to external memory.

To reduce the loss of the controller's IO function to external addressing, several manufacturers multiplex the low order address and data bus to a single eight bit port. This is then labelled P7 through to P0 when in 'single chip mode', i.e. internal memory or D7/A7 through to D0/A0 for external memory mode (Fig. 16.25).

An extra control line AS (Address Strobe) is required to demultiplex lower order addresses from data, or to be precise, to clock the latch when the low order addresses are valid. The latch sustains the lower order addresses freeing up the controller's port to handle the following data cycle. The read/write control may also double up as an IO when in single chip mode.

Fault finding controller based systems in extended address mode are similar to fault finding processor systems, except that the data bus is interleaved with addresses.

When fault finding controllers configured in single-chip-mode, that are not factory masked, you must bear in mind that the internal ROM may not be programmed. The device will still configure itself correctly because its response to the 'mode settings' is built into the hardware. All the IO lines will be set to input (high Z), another hardware response to reset. The controller will then execute the code, i.e. endless FF. How it responds, i.e. whether it cycles round and round its entire address range or whether it branches in a short 'illegal op-code' repetition, is not relevant but it will not be able to program any of its IO lines as an output. That is, if all IOs remain in a high Z condition there is a fundamental problem. It may be that the internal prom is blank, or the reset itself is permanently asserted, or V_{dd} is not connected, or there is no clock, or the device is u/s.

The IO hardware characteristics vary between different controllers. Some have CMOS style outputs and some have open collector style outputs. Others have an open collector output with a weak (70 k–120 k) pullup resistance. Others can program their IO to be either CMOS or open collector.

Example

This nasty production fault occurred in a small control circuit with two microcontrollers. Both were operated in extended address mode, i.e. their fundamental buses and controls were brought out on to the pcb together with their low order address latches. One controller decoded and encoded data and the other one controlled the system.

These boards were exercised and tested by ATE, not the passive 'node to node bed of nails' arrangement, but a custom built machine that exercised the dynamic functioning. All fault finding had to be done on the ATE because it was too time consuming to set up the tests 'by hand'. This meant that we were dependent on the ATE and lacked the direct contact that would have made this fault easier to spot.

When the ATE first applied power it immediately reported excessive current consumption. The 5 V supply rail had dropped to 4 V and one of the controllers was getting hot. Its data bus had contention on all bits so without spending more than a few seconds on it I had it replaced. After replacement, the fault remained but now the other controller was getting hot. After replacing that one the first got hot again; perhaps they were damaging each other (unlikely) so they were both replaced and still the fault remained.

The problem was that the pcb had its own 5 V regulator and the input of the regulator was s/c to the output. The result of this was that the test rig attempted to put 13.8 V into the pcb, the processors naturally crashed, and in crashing they were able to select 'forbidden' IO references and contend the entire data bus (i.e. if a processor tried to write to a ROM for instance). This then made a huge demand on the supply which was current limited via the test rig so that supply voltage was forced to drop and coincidentally fell below 5 V. This was only spotted when the supply was tripped off and on whilst scoping the buses simultaneously. Then, briefly yet

257

clearly, the bus logic high level ran at 5.5 V before immediately falling to 4 V high, with contention at 2 V. This is a fine example. Everyone knows what a logic high level looks like, yet few of us would question that fleeting difference above during switch on (the only clue). I saw it minutes after I powered up the board yet I ignored it, and only after the pcb was virtually destroyed in the ic replacing did I settle down, stop and think.

17 Dual trace oscilloscope

For this type of work the first priority is a times 10 probe. The capacitance of a times one lead together with the input capacitance of the scope filter out some of the delicate effects we should be looking for. The scope itself is secondary, in fact the simpler the better. I would not recommend a digital storage scope, it is better to see the anomaly as it happens rather than spot (or miss) a captured image of it later when fault finding a known working design. (In development work, the reverse of this is true.) So always use the times 10 probe not forgetting that it must be adjusted to cancel the filtering effect of the series resistor and the scope and scope lead's capacitance.

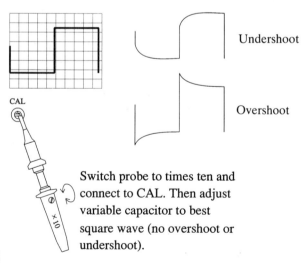

Undershoot

Overshoot

CAL

Switch probe to times ten and connect to CAL. Then adjust variable capacitor to best square wave (no overshoot or undershoot).

Figure 17.1

Dual trace oscilloscope

A regular difficulty when comparing two traces together is that sometimes they refuse to occur simultaneously unless the timebase is tweaked with the variable adjustment. To see why this is, suppose there are two repetitive waveforms that occur simultaneously as in Fig. 17.2. Triggering from the top trace, positive going edges, with the timebase adjusted to view slightly less than one cycle, then the scope would show either the trace in Fig. 17.3(a) or 17.3(b). But in either case both traces are correctly aligned together in time.

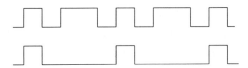

Figure 17.2

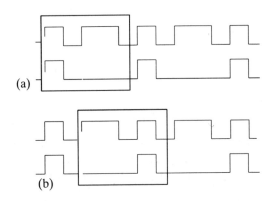

Figure 17.3

If we were to adjust the timebase so that a little more of these waveforms appeared on the scope as in Fig. 17.4, then the scope would show a different picture, not like that in Fig. 17.4 but like Fig. 17.5, which is misleading because the two waveforms are no longer aligned in time: the short pulse does not happen in the middle of the long one. There are now three different possible versions of this picture because there are three different positive edges that the scope could trigger on, but in all three the two traces would be 'misaligned'.

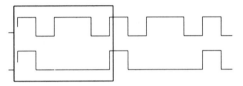

Figure 17.4

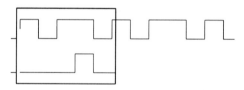

Figure 17.5

The reason is that the scope scans the traces alternately (this problem would not occur in chop mode, but chop mode is unsuitable for high speed work). First the top trace is triggered and the scan begins. When the scan finishes then the next positive edge of the top trace triggers the lower trace and then that scan begins; the scope display is really created as illustrated in Fig. 17.6.

261

Dual trace oscilloscope

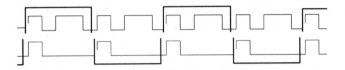

Figure 17.6

To avoid this irritation it is sometimes necessary to adjust the variable timebase pot a little while using the same signal (the triggering one) on both traces until they line up and then you can use any other signal with it and it will always be aligned with the triggering one.

18 Integrated circuit replacement

If replacing an integrated circuit (ic) in the field it is best, if possible, to try and use not only the same technology but also the same manufacturer.

IC removal

Dual inline ics can be removed by:

a Cutting out the body and removing the legs separately, then cleaning out the holes with solder braid or a solder sucker.
b Removing whole with solder braid or solder sucker.
c Removing whole with large hot iron tip and plenty of excess solder then tapping board immediately afterwards to knock off excess solder, leaving clean holes.

The safest method for the novice is (a) which is least likely to damage the pcb. The most difficult is (c), but, for some tough jobs (like removing a 50 pin idc header) obligatory, it must be practised first on a scrap pcb. Use a large Weller no 8 tip, smoothly run the iron back and forth across the underside whilst liberally applying a large bead of fresh solder as the ic (or header) starts to move freely, keep it in place for a few more seconds, then allow it to drop out under its own weight and immediately tap the pcb. I use method (b) mainly; solder braid, if un-tinned and tarnished, is no good – an

American brand called 'Hex Wick' which is clean tinned and fluxed tends to work well (if only half the solder is drawn out of a plated through-hole it must be re-soldered before trying again).

Solder suckers are preferred by most, the trick is to keep the nozzle in contact with the board and lightly touch the tip of the iron; but at the moment of release, the reaction knocks the iron tip into the pcb dislodging any underlying pcb tracking, therefore position solder sucker nozzle/iron tip to keep the tip on glass fibre only, not copper. Immediately the solder sucker is withdrawn push the exposed end of the ic leg way from the hole side breaking the thin remaining solder join. The iron softens the connection allowing movement whilst the rapidly cooling surrounding prevents the joint reforming afterwards when the iron is removed, the whole action takes only 2 seconds. Then turn the pcb right way up and poke the 'shoulders' of the ic legs inwards slightly, they should move easily, some with a slight snap, and the ic should fall out under its own weight.

I keep the solder sucker clean and serviceable with an extra squeeze or two between every joint de-soldered. If it jams whilst doing this then a slight bending of the nozzle while forcing the plunger home will catch the 'tube' of solder at its edge and force the whole lot out in one go, with luck.

This method works well if the pcb has 0.9 mm holes or larger. If smaller it may damage the pth lining if forced and then method (a) must be used.

Surface mount

Surface mount packages are easier to remove/replace than are conventional dil packages if the right tool, a nitrogen gas gun, is available. The inert gas stops the solder oxidizing and 'whiskering' while the remaining flux allows surface tension to 'bead' the solder around each joint. If excessive solder has been introduced so that several or all the ic legs get 'blobbed' together, then a hot iron will pick up that excess, or a slight tap can knock it off. When removing an ic, never pull it way from the pcb, one or more tracks will come off; always push gently from the side, if it moves, keep the heat going for a few more seconds and then remove. I prefer to hold the pcb at 45° with the ic downwards for the last few passes of the gun so that the ic falls away under its own weight.

If the gun is not hot enough then fatigue will induce 'operator error'. A 72 pin package should fall free in under 20 seconds. Burn marks on the pcb will never occur if the gun is always kept moving (especially during approach and removal from the board – the same as spray painting, the work is approached 'on the run' and left in the same way to avoid drips).

If a gas gun is not available, surface mount packages can be removed with expert de-soldering especially if there is a slight gap under each leg, otherwise, if the package is small (14 to 16 pin), then two irons held in either hand can, with extra solder, be run up and down either side of the package until it moves and then deftly flipped clear.

Glossary

Assembler A computer program that converts mnemonics (+ any appended data) of an instruction set into their binary op-codes (+ any appended data).

Back emf The emf generated within an inductor by a moving magnetic field when the moving magnetic field is generated by the same inductor (when a changing current is forced through it).

Bandwidth The size of a range (or band) of frequencies, e.g. an amplifier with a frequency response of 100 MHz to 108 MHz would have a bandwidth of 8 MHz.

CAS Column address strobe; a special signal required by a DRAM to allow it to demultiplex the incoming address (see Chapter 16).

CE Chip enable; often used interchangeably with CS, see pages 155–6.

Character generator A special device that converts a single byte of data into the 'video stream' of a character. The character font is fixed by the CG and cannot be altered. The simple action of the device, a byte of RAM is converted to its equivalent ASCII character on screen, belies its complexity.

Closed loop gain The overall gain of an op-amp after modification by the negative feedback of its supporting components.

CMOS Complementary metal oxide silicon; an integrated circuit technology that requires no power to maintain a logic state and consumes very little power when changing states, i.e. a CMOS device will consume very little power at low clock speeds. This technology also features very high input impedance.

Common emitter A transistor configuration where the output is taken from the collector.

Contention Signifies a collision between output stages sharing a common bus when they are accidentally enabled at the same time.

Crash When a processor is (accidentally) forced to address data (rather than an op-code) during the instruction fetch cycle.

CRT Cathode ray tube; used to display television pictures or oscilloscope traces.

CRTC CRT controller; a specialized ic used in concert with other devices to develop the video display.

CS Chip select; see pages 155–6.

Demultiplex See multiplex below.

DMM Digital multimeter.

DRAM Dynamic random access memory (see Chapter 9 (Memory for RAM) and Chapter 19 (Dynamic RAM)).

DVM Digital voltmeter.

emf Electromotive force; the force that forces a current to flow through a resistance, i.e. a voltage. Emf is the 'pure' original driving voltage that feeds the output impedance. Pd is the actual voltage that exists at the output (after the output impedance), i.e. emf can never be measured directly.

Emitter follower A transistor configuration where the output is taken from the emitter.

EPROM Eraseable programmable read only memory; these can be erased (by exposure to a strong ultraviolet source) and reprogrammed many times.

Feedback Where a portion of the output is fed back to the input. If the feedback opposes the input then it is termed negative feedback. If the feedback enhances the input it is positive feedback.

Frequency response Variation of gain with frequency, e.g. the frequency response of an amplifier is a graph of its output level against a range of input frequencies.

HC High speed CMOS; a CMOS technology that partially conforms to TTL technology, i.e. pinouts and logic functions are copied but CMOS input thresholds are maintained.

HCT High speed CMOS TTL; a CMOS technology that conforms to TTL technology as far as pinouts, logic functions and input thresholds are concerned.

Impedance Resistance plus reactance, i.e. at dc a thing will exhibit fixed x ohms of resistance but at ac it will also exhibit a (varying with frequency) y ohms of reactance which is added to the fixed resistance to give the total impedance (Z).

Inductor A device designed to store energy as a magnetic field.

irq Interrupt ReQuest; an 'asynchronous' input to a processor, i.e. processors will follow their fixed paths and branches through their programs in strict sequential fashion. The irq input provides a way to interrupt this fixed sequence 'at any time' to 'force a faster response' (see page 183).

lsb Least significant bit; i.e. bit 0 (D0).

Microcontroller See pages 254–6.

Mnemonic A reminder, a memory aid.

MOS Metal oxide silicon; an integrated circuit technology featuring high input impedance.

MPU Microprocessor unit; this is abbreviated to 'processor' throughout this book, i.e. it is the main subject of Part 2.

MSB Most significant bit; i.e. in an eight bit system, bit 7 (D7).

Multiplexer Multiplexing is the technique of switching two (or more) signals onto a single wire (at the transmit end). At the receive end, the signals are demultiplexed, i.e. switched from the single wire back to reproduce the two (or more) original signals.

MUX Multiplexer; see above.

Negative feedback See Feedback above.

NPN N-type/P-type/N-type; symbolic of a Transistor type whose base should be positive wrt its emitter in normal operation.

Open loop gain The gain of an op-amp (without any feedback).

Operating system A program that controls the hardware of a system.

pcb Printed circuit board.

pd Potential difference; the voltage between two points.

PNP P-type/N-type/P-type; symbolic of a transistor type whose base should be negative wrt its emitter in normal operation.

Positive feedback See Feedback above.

PROM Programmable read only memory; the older type of ROM which could be programmed (once only) by the end user. Today's PROMs are OTPROMs (One Time PROM) which are eproms without the quartz window.

Q Quality; is a measure of the 'quality' of a tuned circuit. If it were perfect, i.e. if there were no resistance but only a pure inductance and a pure capacitance then it would be able to 'ring' forever and the Q would be infinite (see page 50).

RAM Random access memory; see Chapter 9 (Memory).

RAS Row address strobe; a special signal required by a DRAM to allow it to demultiplex the incoming address (see Chapter 16).

Rectifier Diode(s) used in converting ac to dc.

RF Radio frequency.

ROM Read only memory; see Chapter 9 (Memory).

Scope Abbreviation of oscilloscope.

UUT Unit under test.

V_{dd} The positive supply pin of a MOS or CMOS ic.

VDU Visual display unit; i.e. the monitor of a terminal or computer.

V_{ss} The negative (ground) supply pin of a MOS or CMOS ic.

Index